Pytest 企业级应用实战

温红化 编著

北京航空航天大学出版社

内 容 简 介

Pytest 是非常广泛的基于 Python 语言的应用自动化测试框架,本书首先从实战的角度层层递进地讲解了 Pytest 框架使用方法,然后从企业级应用实战的角度讲解了如何利用 Pytest 自动化测试企业测试框架,并结合 Jenkins 以及 Allure 工具,演示在企业中是如何应用的。

第 1 章介绍运行环境以及 Pytest 快速体验,第 2 章介绍 Pytest 测试脚本的组成结构和测试脚本默认的命名规则以及脚本执行,第 3 章介绍断言的使用方法,第 4 章介绍了标签的使用方法,第 5 章详细介绍了自动化脚本各种常用的执行策略,第 6~8 章介绍了 fixture 的基础和高级应用以及常见的 fixture 的使用方法,第 9 章介绍了参数化以及数据驱动的使用方法,第 10 章介绍了告警,第 11 章介绍了 Pytest 中丰富的日志和打印功能,第 12 章介绍了 Allure 测试报告,第 13 章从设计角度介绍如何设计封装自动化测试框架,以及 Pytest、Allure 与 Jenkins 的集成,第 14 章介绍了常用的第三方插件,第 15 章解读了 Pytest 的核心即 Pluggy 的源码分析。

本书既可以作为自动化测试人员、测试开发人员的工具书,也可以作为 Pytest 技术相关培训教材。

图书在版编目(CIP)数据

Pytest 企业级应用实战 / 温红化编著. -- 北京:
北京航空航天大学出版社,2023.7
ISBN 978-7-5124-4116-3

Ⅰ. ①P… Ⅱ. ①温… Ⅲ. ①软件工具—程序设计
Ⅳ. ①TP311.561

中国国家版本馆 CIP 数据核字(2023)第 125417 号

版权所有,侵权必究。

Pytest 企业级应用实战
温红化 编著
策划编辑 杨晓方 责任编辑 杨晓方

*

北京航空航天大学出版社出版发行

北京市海淀区学院路 37 号(邮编 100191) http://www.buaapress.com.cn
发行部电话:(010)82317024 传真:(010)82328026
读者信箱:copyrights@buaacm.com.cn 邮购电话:(010)82316936
涿州市新华印刷有限公司印装 各地书店经销

*

开本:710×1 000 1/16 印张:24 字数:540 千字
2023 年 7 月第 1 版 2023 年 7 月第 1 次印刷
ISBN 978-7-5124-4116-3 定价:99.00 元

若本书有倒页、脱页、缺页等印装质量问题,请与本社发行部联系调换。联系电话:(010)82317024

前 言

目前，软件自动化测试在整个IT行业或者说在软件测试领域正占据越来越重要的位置，当前整个IT行业都在提倡降本增效，而自动化测试则是降本增效的一个非常重要的手段。此外，从个人成长角度来看，测试开发将是未来测试人员的职业发展趋势，而自动化测试是测试开发最容易入手的方向。那么想要做好自动化测试，至少需要对一种自动化测试框架非常熟悉，在众多自动化测试框架中，Pytest可以说是独领风骚。本书从实战的角度，详细介绍Pytest自动化测试框架的使用方法，同时融入了一些在企业中的实战应用经验。

本书共分15章，其中第1章主要内容为快速体验Pytest，第2～11章让读者循序渐进地了解Pytest，第12～14章则从设计自动化测试框架的角度介绍Pytest，第15章从源码的角度介绍了Pytest自动化测试框架的核心原理。

为了更好地满足技术类读者的需求，本书不过多地介绍理论，而是采用实例演示的方式，与读者一同学习和总结Pytest的用法，进而能让读者达到更好的学习效果。此外，本书还可以作为查询工具书，书中各章节标题采用解决实际问题类型的方式来命名，以便读者可以迅速定位到自己喜欢的或需要的章节进行阅读。

本书的特点如下：

（1）以实战为主，尽量减少过多文字理论层面的分析，展示实战代码，即从实战中总结出Pytest的使用方法，通过引入实际问题激发读者深入分析并解决问题的兴趣。

（2）章节安排上循序渐进，先让读者快速体验Pytest，然后逐渐介绍Pytest各个方面的特性以及使用方法，接下来介绍在企业级实战中如何将Pytest与其他工具结合，从而设计出自动化测试团队真正需要的实际可执行、可落地的自动化测试框架，最后又详细解读Pytest最核心的源码，进而满足了一些进阶学习读者的需求。

（3）本书在介绍Pytest各个方面特性和使用方法的内容中，既有基础内容的讲解，也有一些进阶内容，让读者既能迅速地掌握Pytest的基础内容，同时也能很快地体会学习Pytest带来的成就感。

（4）设置大量的代码示例，让读者通过实战来总结和研究Pytest的用法，这也是学习技术最佳的途径。

由于编者水平有限，书中难免有不成熟和错误的地方，在这里恳请读者批评指正。读者反馈发现的问题、获取相关资料或有实验平台技术问题，均可发信至邮箱：hitredrose@163.com进行咨询。

编 者

目 录

第1章 Pytest 概述 ... 1

1.1 Pytest 简介 ... 1
- 1.1.1 Pytest 的特点 ... 1
- 1.1.2 Pytest 发展历程 ... 1
- 1.1.3 为什么要选用 Pytest ... 2

1.2 开发环境的安装 ... 3
- 1.2.1 Python 安装 ... 3
- 1.2.2 同时安装多个 Python 版本 ... 8
- 1.2.3 彻底卸载 Python ... 9
- 1.2.4 Pycharm 的下载安装 ... 10

1.3 虚拟环境管理工具 Pipenv 的应用 ... 15
- 1.3.1 安装 Pipenv ... 16
- 1.3.2 Pipenv 工具的使用方法 ... 17
- 1.3.3 配置 Python 解释器 ... 20

1.4 Pytest 快速体验 ... 23
- 1.4.1 Pytest 安装与升级 ... 23
- 1.4.2 创建并执行第一个测试脚本 ... 23
- 1.4.3 Pytest 命令的默认行为 ... 24
- 1.4.4 对产生的异常进行断言 ... 24
- 1.4.5 在类中编写测试脚本 ... 25
- 1.4.6 测试脚本请求,创建一个临时目录 ... 28

第2章 Pytest 脚本规则、组成与运行 ... 29

2.1 Pytest 脚本的规则 ... 29
- 2.1.1 测试脚本文件命名规则 ... 29
- 2.1.2 测试函数的测试类命名规则 ... 30
- 2.1.3 测试脚本目录的约束条件 ... 32
- 2.1.4 自定义测试文件名、测试类、测试函数命名规则 ... 34

2.2 Pytest 脚本的组成 ... 35
- 2.2.1 自动化脚本组成简介 ... 35
- 2.2.2 测试类中各个层级的 setup 和 teardown ... 36

  2.2.3 测试文件中各个层级的 setup 和 teardown ·················· 37

  2.2.4 测试文件中测试函数和类中测试方法混合时各个层级的 setup 与
     teardown ·· 39

  2.2.5 各个层级的 setup 和 teardown 在自动化实践中的应用 ········ 40

2.3 Pytest 脚本的运行 ·· 41

  2.3.1 指定目录或文件 ·· 41

  2.3.2 指定测试函数或测试方法 ·· 41

  2.3.3 通过--k 参数对文件类名及函数名进行模糊匹配和挑选 ········ 43

  2.3.4 通过--ignore 参数挑选用例,忽略执行 ·························· 46

  2.3.5 通过--ignore-glob 参数支持正则挑选用例忽略 ················ 47

  2.3.6 通过--deselect 参数挑选用例不执行,并显示未执行数量 ······ 48

  2.3.7 通过重复指定文件路径,实现用例重复执行 ···················· 49

  2.3.8 通过--collect-only 参数不执行脚本而统计脚本总数 ············ 50

  2.3.9 通过 Pytest.ini 设置用例默认的搜索目录 ······················· 52

  2.3.10 在 IDE 中通过右键执行当前文件用例 ························· 53

2.4 Pytest 脚本的加载原理 ·· 54

  2.4.1 prepend 模式 ··· 54

  2.4.2 append 模式 ·· 56

  2.4.3 prepend 和 append 模式存在的问题 ····························· 57

  2.4.4 importlib 模式 ··· 59

第 3 章 Assert 断言 ·· 61

3.1 使用 Assert 断言 ·· 61

  3.1.1 Python 中为 False 的数值断言均失败 ··························· 61

  3.1.2 Python 逻辑表达式为 False 的断言均失败 ····················· 63

3.2 自定义断言报错信息 ··· 65

3.3 对捕获的异常进行断言 ·· 66

  3.3.1 对异常类型进行断言 ··· 66

  3.3.2 对捕获的异常信息进行断言 ···································· 68

  3.3.3 同时对捕获的异常类型和异常信息进行断言 ················ 69

  3.3.4 对一个函数可能产生的异常进行断言 ························ 70

3.4 重写断言 Assert 语句的报错信息 ··································· 71

  3.4.1 默认的报错信息 ··· 71

  3.4.2 重写判断是否相等的断言报错信息 ···························· 72

  3.4.3 重写常见的判断逻辑报错信息 ·································· 74

第 4 章　mark 标签的用法 ·················· 80

4.1　skip 和 skipif 的使用方法 ············· 80
4.1.1　skip 的用法 ················· 80
4.1.2　skipif 的用法 ················ 83

4.2　xfail 和 xpass 的用法 ··············· 85
4.2.1　xfail 标记测试脚本 ·············· 85
4.2.2　xfail 根据条件判断标记测试脚本 ········ 86
4.2.3　动态启用 xfail 标记 ············· 87
4.2.4　@pytest.mark.xfail 只设置 reason 参数 ···· 89
4.2.5　@pytest.mark.xfail 通过 run 参数设置是否执行 ·· 89
4.2.6　xpassed 用例显示为失败 ··········· 90
4.2.7　使 xfail 标记失效的方法 ··········· 91

4.3　importorskip 的用法 ··············· 92

4.4　注册并使用自定义 mark 标签 ············ 94
4.4.1　直接使用自定义 mark 标签 ·········· 94
4.4.2　通过 conftest.py 文件重写 pytest_configure 函数的注册标签 ··· 96
4.4.3　通过 pytest.ini 文件配置注册标签 ······· 97
4.4.4　通过标签灵活挑选测试脚本执行 ········ 98

第 5 章　Pytest 测试用例的执行策略 ············ 101

5.1　在遇到用例失败时如何停止执行 ·········· 101
5.2　如何在用例失败时打印局部变量 ·········· 102
5.3　如何在用例执行失败时使用 pdb 进行调试 ······ 105
5.4　用例失败后如何重新执行 ············· 107
5.5　如何在一个用例断言失败后继续执行 ········ 110
5.6　如何在失败 N 个用例后停止执行 ·········· 113
5.7　如何只执行上次失败的用例 ············ 115
5.8　如何从上次失败处继续执行用例 ·········· 119
5.9　如何先执行上次失败用例，再执行其他用例 ····· 123
5.10　如何重复执行用例 ··············· 125
5.11　如何进行多进程并行执行用例 ·········· 130
5.12　如何随机执行用例 ··············· 133
5.13　如何只运行未提交 git 代码仓的脚本 ········ 138
5.14　如何查找耗时最长的用例脚本 ·········· 141

第 6 章 fixture 的基础应用 · 148

- 6.1 fixture 传值的作用 · 148
- 6.2 fixture 嵌套的应用 · 149
- 6.3 在函数中调用多个 fixture · 151
- 6.4 fixture 如何设置自动执行 · 152
- 6.5 通过 yield 实现 setup 和 teardown 的功能 · 154
- 6.6 function 级别的 fixture · 157
- 6.7 class 级别的 fixture · 159
- 6.8 module 级别的 fixture · 161
- 6.9 package 级别的 fixture · 166
- 6.10 session 级别的 fixture · 175
- 6.11 fixture 的覆盖原则 · 177
- 6.12 yield 的缺陷及解决方案 · 181

第 7 章 fixture 的高级应用 · 186

- 7.1 通过 request 动态获取或配置测试脚本的属性 · 186
- 7.2 通过 request 向 fixture 传递参数 · 190
- 7.3 fixture 如何实现参数化,即数据驱动 · 193
- 7.4 fixture 参数化指定用例 id · 197
- 7.5 fixture 参数化中指定参数使用 skip 标记 · 198
- 7.6 fixture 参数化时,指定参数使用 xfail 标记 · 200
- 7.7 fixture 参数化可实现两组数据的全排列组合测试 · 202
- 7.8 通过 usefixtures 为一个测试类调用 fixture · 209

第 8 章 常见内置 fixture 的应用 · 212

- 8.1 如何进行文档测试 · 212
- 8.2 如何使用猴子补丁进行异常测试 · 215
- 8.3 如何使测试过程中产生的文件自动删除 · 218
- 8.4 如何动态获取 Pytest.ini 中的配置以及命令行参数 · 221
- 8.5 如何在运行中动态获取用例的属性 · 223

第 9 章 parameterize 参数化及数据驱动 · 225

- 9.1 测试函数使用 parametrize 进行参数化 · 225
- 9.2 测试类使用 parametrize 进行参数化 · 227
- 9.3 通过 pytestmark 对测试模块内的代码进行参数化 · 228
- 9.4 parametrize 参数化时使用 skip 标记 · 230

9.5　parametrize 参数化时使用 xfail 标记 ……………………………… 231

9.6　parametrize 参数化时对两组数据进行全排列组合测试 ……………… 232

第 10 章　告　警　239

10.1　如何使用命令行控制告警 ……………………………………………… 239

10.2　如何通过 filterwarnings 配置告警或将告警报错 …………………… 244

10.3　如何将一个测试文件产生的告警都忽略或者转换为报错 …………… 249

10.4　如何关闭所有告警显示 ………………………………………………… 253

10.5　如何通过 Pytest.ini 配置告警或将告警报错 ………………………… 255

10.6　如何对产生的告警进行断言 …………………………………………… 259

10.7　如何通过 recwarn 记录用例中产生的告警 ………………………… 262

第 11 章　日志和控制台输出管理　264

11.1　实时标准输出和捕获标准输出 ………………………………………… 264

11.2　如何打开或关闭实时输出和捕获标准输出 …………………………… 266

11.3　如何使用 logging 模块写日志 ………………………………………… 268

11.4　什么是实时日志和捕获日志 …………………………………………… 272

11.5　如何打开或关闭实时日志和捕获日志 ………………………………… 274

11.6　caplog 的应用场景及使用方法 ………………………………………… 280

　　11.6.1　如何在测试用例中设置日志级别 ………………………………… 280

　　11.6.2　如何对日志级别进行断言 ………………………………………… 282

　　11.6.3　如何对日志内容进行断言 ………………………………………… 283

　　11.6.4　如何对日志级别和日志内容同时进行断言 ……………………… 284

　　11.6.5　在测试用例中如何获取 setup 中的日志 ………………………… 285

11.7　Pytest 如何进行正确配置及使用日志功能 …………………………… 287

第 12 章　Allure 测试报告　289

12.1　Windows10 安装 Allure ………………………………………………… 289

12.2　使用 Allure 生成测试报告 ……………………………………………… 292

12.3　定制 Pytest 自动化测试报告样式 ……………………………………… 296

12.4　定制 Allure 报告中的 logo ……………………………………………… 299

第 13 章　与 jenkins 持续集成　303

13.1　设计开发自动化测试框架 ……………………………………………… 303

13.2　测试脚本上传 git 代码仓库 …………………………………………… 309

13.3　使用 docker 搭建 jenkins ……………………………………………… 314

13.4　在 Centos7 系统中安装 git 客户端 …………………………………… 318

13.5 在 Centos7 系统中安装配置 Allure ……………………………… 319
13.6 Jenkins 基础配置 …………………………………………………… 321
　　13.6.1 为 Jenkins 增加节点 ……………………………………… 321
　　13.6.2 为 Jenkins 配置 git 和 Allure 工具位置 ………………… 324
13.7 基于 Jenkins 创建构建任务并生成 Allure 报告 ………………… 328

第 14 章 其他常用的用例插件 ……………………………………… 335
14.1 Pytest-attrib 根据属性挑选用例 ………………………………… 335
14.2 Pytest-sugar 执行过程中显示进度条 …………………………… 338
14.3 Pytest-csv 执行结果输出 csv 文件 ……………………………… 339
14.4 用 Pytest-tldr 插件简化脚本执行日志输出 …………………… 340

第 15 章 Pytest 核心即 Pluggy 源码解读 ………………………… 344
15.1 Pluggy 模块的应用 ………………………………………………… 344
15.2 Pluggy 源码解读基础准备 ………………………………………… 351
15.3 HookspecMarker 类和 HookimplMarker 类分析 ……………… 354
15.4 如何将 PluginManager 类实例化 ………………………………… 358
15.5 为 add_hookspecs 增加自定义的接口类 ………………………… 362
15.6 register 注册插件源码解析 ……………………………………… 364
15.7 hook 函数调用执行过程分析 …………………………………… 368
15.8 PluginManager 类的其他功能 …………………………………… 370

第 1 章 Pytest 概述

1.1 Pytest 简介

1.1.1 Pytest 的特点

Pytest 是一款基于 Python 开源的自动化测试框架，具体来说，它有以下特点：
- 使用简单，非常容易上手。
- 功能强大，可对大部分测试活动自动化。
- 开源生态健康，文档丰富。
- 插件化架构，可扩展性强。
- 拥有非常丰富的插件库，可覆盖现有的大部分自动化需求。
- 支持多种执行方式，执行调度灵活。
- 支持参数化，支持数据驱动测试。
- 能够兼容其他自动化测试框架编写的脚本。
- 结合 Allure 插件可生成非常漂亮的测试报告。
- 方便与 Jenkins 进行持续集成。

1.1.2 Pytest 发展历程

在正式学习 Pytest 之前，我们先大概了解一下 Pytest 的发展历程。Pytest 从 2009 年发布了 1.0.0 版本后，不断迭代更新，到 2022 年已经更新到 7.0.0 版本了。Pytest 的演变过程如下：
- 2009 年 8 月发布了 1.0.0 版本。
- 2010 年 11 月发布了 2.0.0 版本。
- 2016 年 8 月发布了 3.0.0 版本。
- 2018 年 11 月发布了 4.0.0 版本。
- 2019 年 6 月发布了 5.0.0 版本。
- 2020 年 7 月发布了 6.0.0 版本。
- 2022 年 2 月发布了 7.0.0 版本。

1.1.3 为什么要选用 Pytest

自动化测试框架很多，选择合适的自动化测试框架是非常重要的。在选择自动化测试框架之前，首先要选择自动化测试框架的语言，当前，对于测试，Python 语言是最优选，原因就是 Python 语言简单易上手，学习成本小。此外 Python 社区非常活跃，Python 遇到的问题在社区基本都能找到答案，当然 Python 语言也是当下非常火的语言，但 Python 也是有缺点的，如 Python 运行速度相对比较慢，这一点在应用程序或底层程序中影响是很大的。但是对于自动化测试而言，这点影响可以忽略，因为自动化测试脚本的目标是将测试用例自动化，提高测试用例执行效率。至于运行速度，自动化测试没有严格的要求。因而，对自动化测试而言，Python 语言是最优选。

选定了 Python 语言，就可以在基于 Python 语言的自动化框架中进行选择，而基于 Python 的自动化测试框架，相对比较流行的就三个，即 unittest、RobotFramework、Pytest，下面首先对比一下这三款自动化测试框架的优缺点。

1. unittest 的优点

- Python 内置框架，无须安装，不存在是否兼容问题。
- 简单，易上手，学习成本低。
- 适合于 Python 项目的单元测试场景。

unitetst 的缺点是功能较少，在应用系统测试中则很难满足测试的较多需求。

2. RobotFramework 的优点

- 使用表格形式编写脚本，语法简单，学习成本低，易上手。
- 使用关键字驱动，自动化开发简单而便捷。
- 支持中文，适合测试团队全员自动化流程。
- 开源，提供许多第三方库。
- 提供分类和标记，可自由选择执行用例。
- 基于 Python 语言，用户可以自定义 Python 库。

3. RobotFramework 的缺点

- 使用表格形式写脚本，不支持复杂语法，不够灵活。
- 兼容性较差，当一个依赖库发生版本变化，容易出现依赖的问题。
- 近些年，社区不太活跃，特别是随着 Python 升为 3.6 后，大量使用 RobotFramework 的仍然停留在 Python2.7 的基础上。

Pytest 的优点即同在上述 1.1.1 节中提到的 Pytest 的特点是一致的，Pytest 的缺点就是对脚本开发的代码能力有一定要求，即脚本开发人员要有一定的 Python 语言功底，使用起来才方便。

综上所述，unittest 适合基于 Python 语言的项目中的白盒单元测试。对测试报告等要求不高，RobotFramework 适合于测试团队整体代码能力相对较弱的情况，而且项目只考虑对功能测试实现自动化的情况，若随着项目的发展，项目需要对性能测试压力测试等，RobotFramework 的弱点就逐渐显现出来。此外对个人成长而言，每个人都希望能在自动化测试工作中不断提升自身的代码能力，RobotFramework 却不能提供这样的机会，而 Pytest 则非常强大。在项目起始阶段，可以使用 Pytest 的简单的功能，即 Pytest 上手很容易，随着项目的发展壮大以及个人代码能力的提升，可以使用 Pytest 的高级功能，因为 Pytest 是完全基于 Python 语言的，压力测试、性能测试等也是很容易实现。因此，在对项目进行自动化测试框架选型的时候，Pytest 是非常不错的选择。

1.2　开发环境的安装

1.2.1　Python 安装

对于 Python，推荐大家使用 Python3.6 及以后的版本，如果已经安装过 Python3.6 或以后的版本了，则直接使用即可，不需要重新安装 Python 环境，如果尚未安装 Python 环境，则可以按照如下步骤下载安装 Python 软件：

步骤 1：打开 Python 官网，单击【downloads】→【all releases】，如图 1-1 所示。

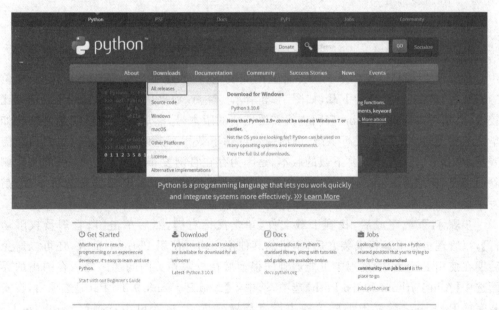

图 1-1　Python 官网

步骤 2：如图 1-2 所示，这里展示了 Python 各个版本的当前状态以及最后支持的时间节点，可以看出当前 Python3.10 处于修复 bug 的状态，Python3.7、Python3.8 和 Python3.9 均已经处于稳定的状态了，此外，Python3.9 支持到 2025 年 10 月，因此，这里建议选择 Python3.9 的版本。然后可以选择 Python3.9 版本中最新的版本

Python3.9.13，大家可单击【Download】链接。

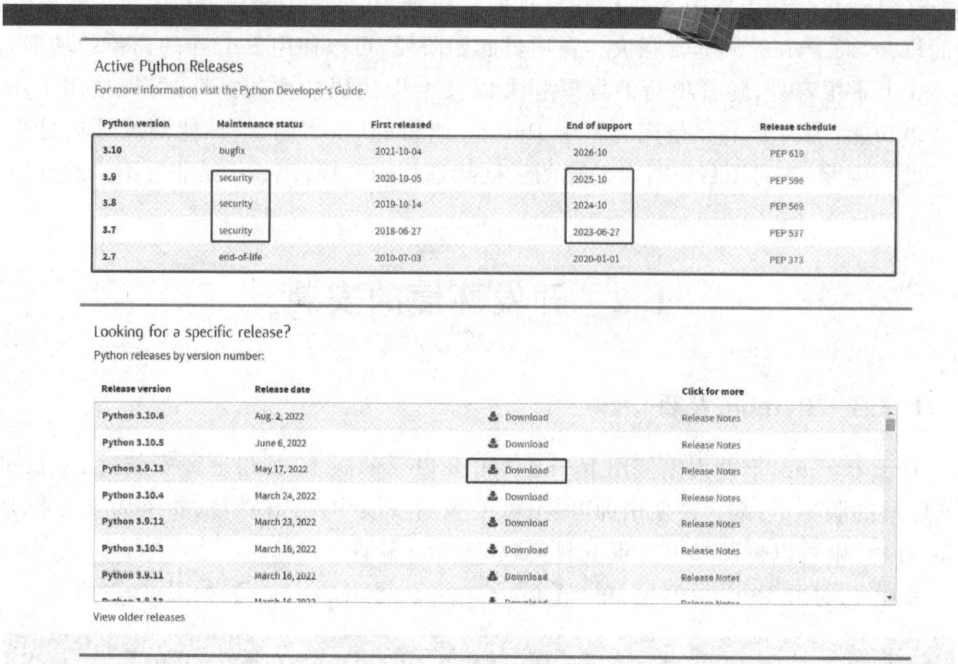

图1-2 Python版本及下载

步骤3：根据具体操作系统选择合适的安装包，比如这里选择 windows 版的 64 位的安装包，单击链接，开始下载，如图1-3所示。注意，如果从 Python 官网下载速度比较慢，大家可以通过 Python 官网查找需要安装的版本号，然后在国内镜像源网站比如华为云找到对应的版本进行下载，国内镜像网站下载相对速度比较快。这里之所以需要从 Python 官网查找需要下载的版本，是因为存在官网发布了版本以后，可能会有删除的情况，比如像 Python3.8.13 的版本，官网当前就没有可供下载的安装版本，但是发布页面却存在3.8.13的版本发布记录。

步骤4：下载完成后，找到下载文件，单击鼠标右键，然后单击【以管理员权限运行】，以管理员权限运行安装之后，本机的其他用户均可使用 Python，此外，还可以避免后期在使用 Python 的过程中可能遇到的很多潜在权限相关的问题。之后在弹出的界面选中【Add Python3.9 To Path】选项，选中了【Add Python3.9 To Path】选项后，在安装完成 Python 之后就不需要单独去设置环境变量了，接着单击选择【Customize Install】，如图1-4所示，即选择自定义安装的方式。如果选择【Install Now】则会默认安装到系统 C 盘，因此建议选择自定义安装，这样就可以在后面的步骤中将安装位置选择 D 盘，否则很容易将 C 盘空间占满，而 C 盘空间释放又相对比较麻烦。

步骤5：此时保持默认选项，即默认安装 pip 工具，同样默认为本机所有用户安装，然后单击【next】按钮，如图1-5所示：

步骤6：如图1-6所示，选中【Install for all users】选项和【Precompile standard li-

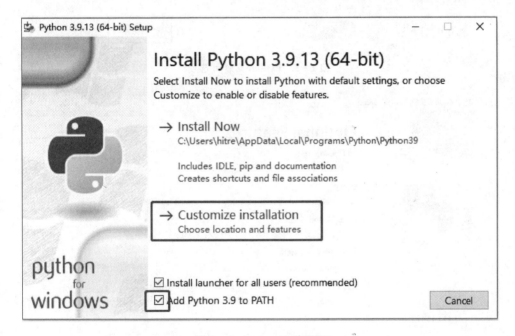

图 1-3　Python 安装包下载

图 1-4　Python 安装界面

brary】选项,然后单击【Browse】按钮,在本地磁盘,比如 D 盘根目录创建 Python39 文件夹,然后设置安装路径为 D:/Python39,这里需要注意,建议直接在磁盘根目录创建,方便后续使用,此外,安装目录建议使用 Python 加版本号的方式,比如安装 Python3.9 就使用 Python39 的目录,如果安装 Python3.8 则使用 Python38 的目录,这样对后续同时安装并使用多个 Python 版本的时候是非常方便的。设置完成后,单击【Install】按钮。

步骤 7:开始安装,此过程可能需要大约 1 min 的时间,如图 1-7 所示。

图 1-5 Python 安装可选项

图 1-6 Python 安装高级选项

步骤 8：如图 1-8 所示，此时 Python 已经安装完成了，这里还有一个选项可根据实际情况而定，即【Disable path length limit】。早在 Window 7 系统时，系统默认的文件路径最大长度为 260 个字符，到 Window10 后，系统仍然沿用之前的配置，即限制 Windows 上文件路径的最大长度为 260 个字符，但是在平时的工作中，总会遇到目录层次结构比较深而导致文件路径的最大长度超过 260。Windows10 系统支持通过手工修改此限制，这里在安装 Python 的时候提供了这个便利，可以再次单击进行修改。这里推荐单击

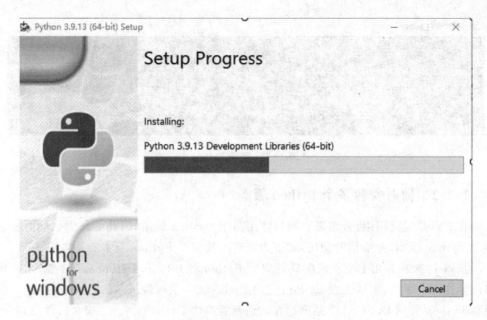

图 1-7　Python 安装进度

修改,即放开文件最大长度 260 的限制。然后单击【close】即完成 Python 的安装。

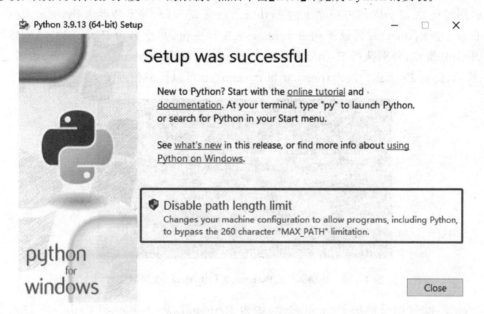

图 1-8　Python 安装完成

步骤 9：如图 1-9 所示,打开 cmd 窗口,执行 Python 命令,回显能正确地显示出 Python 的版本号,表示此时 Python 已经正确地完成安装。

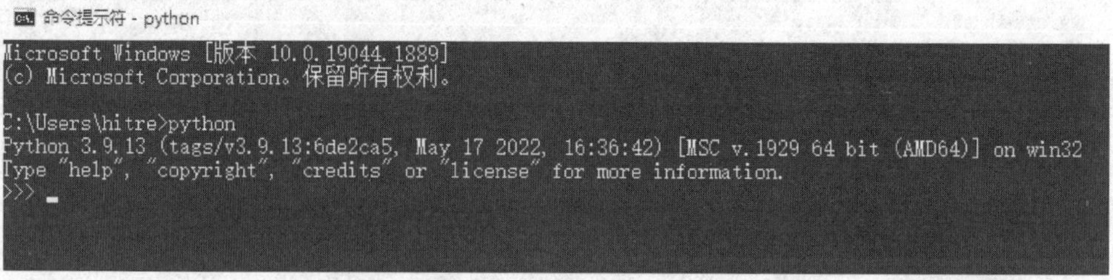

图 1-9　验证 Python

1.2.2　同时安装多个 Python 版本

在实际开发应用中，常常需要同时使用多个 Python 版本，因此必须学会同时安装多个 Python 版本，并能同时使用，如下为同时安装多个 Python 版本的具体步骤。

步骤 1：参照 1.2.1 小节依次进行安装 Python3.8.10 和 Python3.7.9 版本，之所以选择这两个版本，是因为在 Python 官网 Python3.8 系列最新存在的版本为 3.8.10，而在 Python 官网 Python3.7 系列存在的最新版本为 Python3.7.9。安装目录分别为：D:/Python38 和 D:/Python37。

步骤 2：将 D:/Python37/目录下的 Python.exe 复制一份并重命名为 Python37.exe，同理将 D:/Python38/目录下的 Python.exe 复制一份，并重命名为 Python38.exe，将目录 D:/Python39/目录下的 Python.exe 复制一份，并重名为 Python39.exe，然后打开 cmd 窗口，分别执行 Python37-V，Python38-V 和 Python39-V 命令，如图 1-10 所示，此时 Python3.7、Python3.8 和 Python3.9 可以同时使用了。

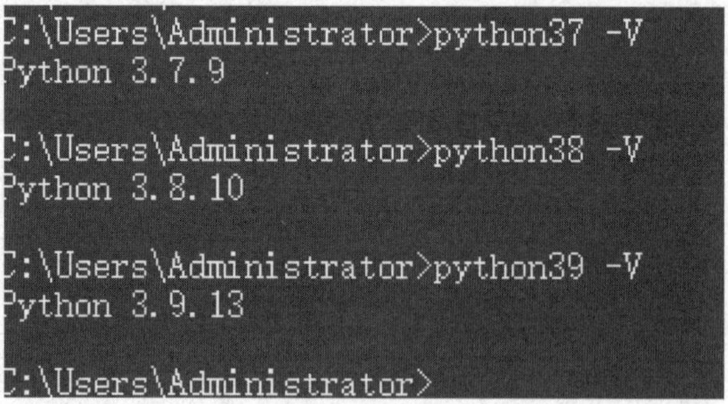

图 1-10　Python3.7、Python3.8 和 Python3.9 同时生效

注意，此时如果使用 Python 命令，因为 Python37、Python38 和 Python39 目录下均有 Python.exe，所以此时 Python 命令指的是在系统环境变量 Path 中 D:/Python37/、D:/Python38/和 D:/Python39 哪个在第一个位置，就是指哪一个，比如这里 D:/Python37/在 Path 中最前面，因此此时 Python-V 显示的就是 Python3.7 版本。另外，此时 Pip 命令的版本和 Python 命令对应的 Python 版本是同一个版本，而如果使用对应环境的 Pip 命令，就应该使用 Pip3.7、Pip3.8 和 Pip3.9，如图 1-11 所示。

```
C:\Users\Administrator>python -V
Python 3.7.9

C:\Users\Administrator>pip -V
pip 20.1.1 from d:\python37\lib\site-packages\pip (python 3.7)

C:\Users\Administrator>pip3.7 -V
pip 20.1.1 from d:\python37\lib\site-packages\pip (python 3.7)

C:\Users\Administrator>pip3.8 -V
pip 21.1.1 from d:\python38\lib\site-packages\pip (python 3.8)

C:\Users\Administrator>pip3.9 -V
pip 22.0.4 from D:\Python39\lib\site-packages\pip (python 3.9)

C:\Users\Administrator>
```

图 1-11 Python 默认版本以及 pip 的使用方式

此外,需要注意,在 D:Python37/、D:Python38/和 D:Python39/目录中,一定要复制,并重命名 Python.exe,如果直接将 Python.exe 重命名,则 Pip3.7、Pip3.8 和 Pip3.9 的命令将无法使用,所以需要格外注意。

1.2.3 彻底卸载 Python

在安装 Python 的过程中,难免会出现比如安装错误或想替换掉旧的版本等情况,所以就需要彻底卸载 Python。以下为彻底卸载 Python 的步骤。

步骤 1:打开控制面板,单击【程序和功能】,如图 1-12 所示。

图 1-12 控制面板

步骤 2:这里以卸载 Python3.7.9 为例,找到 Python3.7.9,然后在其上用右键单击【卸载】,如图 1-13 所示,即开始卸载。

步骤 3:卸载完成后,找到安装目录 D:/Python37,然后将目录 D:Python37 彻底删除,即完成对 Python3.7.9 的彻底卸载。此时,可以打开 cmd 窗口,执行 Python.37-V,如图 1-14 所示,Python3.7.9 已经卸载干净。

图 1-13 卸载 Python3.7.9

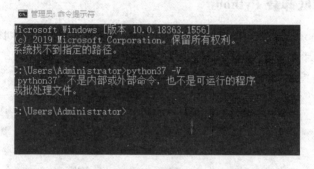

图 1-14 查看 Python3.7.9 界面

1.2.4 Pycharm 的下载安装

编写 Python 代码推荐大家使用 Pycharm 工具，一方面原因是目前 Python 代码开发人员使用 Pycharm 工具的比例比较大，方便与其他人协作。另一方面是 Jetbrain 系列的 IDE 工具覆盖了当前 Java、Python、C、C++、Golang 等语言，使用 Pycharm 便于在开发其他语言时使用统一的 IDE 工具，这样可以保持相同的使用习惯以及类似的快捷键。下面就是 Pycharm 的下载安装步骤。

步骤1：打开 Pycharm 的官网，然后单击【下载】，如图 1-15 所示。

步骤2：这里根据自己的操作系统选择合适的版本，其中又分为专业版和社区版，其中，社区版是免费的，专业版是收费的，如果没有购买，则可以选择社区版，这里以下载专业版为例，如图 1-16 所示，单击【下载】按钮，即开始下载。

步骤3：下载完成后，找到下载的安装包文件，然后用鼠标在其上右键单击【以管理

图 1-15　Pycharm 官网

图 1-16　Pycharm 下载

员权限运行】,如图 1-17 所示,单击【Next】按钮。

步骤 4:通过单击【Browser】按钮,更改安装位置,安装软件尽量不要直接安装在 C 盘,否则很容易导致 C 盘磁盘占满,比如这里选择安装在 D 盘,如图 1-18 所示,单击【Next】按钮,继续安装。

步骤 5:如图 1-19 所示,全部选中列表,然后单击【Next】按钮,继续安装。

步骤 6:如图 1-20 所示,单击【Install】按钮,开始安装。

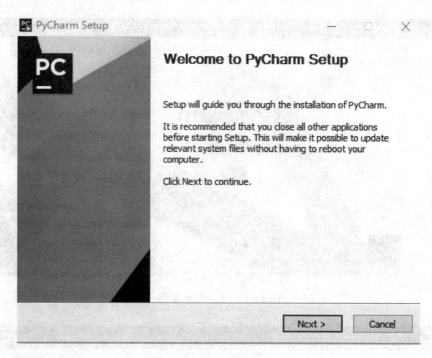

图 1-17 Pycharm 安装向导

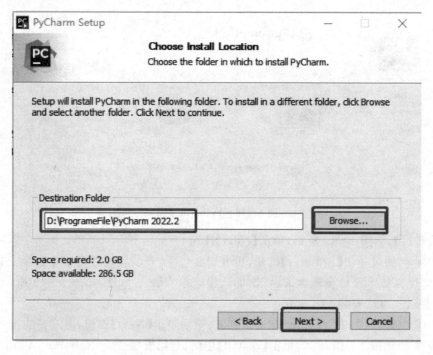

图 1-18 选择安装位置

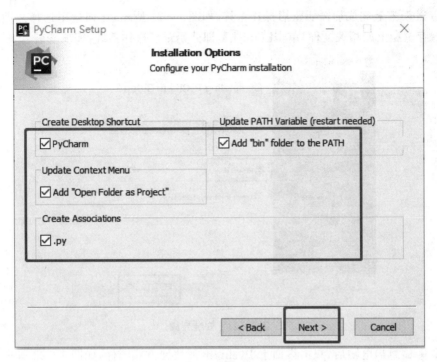

图 1 – 19　设置安装选项

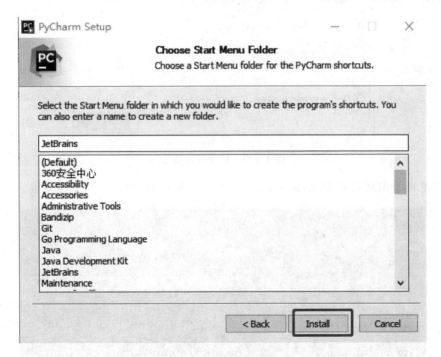

图 1 – 20　开始安装

步骤 7：安装完成后需要重启操作系统，如图 1-21 所示，比如这里选择立即重启，注意保存电脑上的相关文件，单击【Finish】，即开始重启操作系统，完成安装。

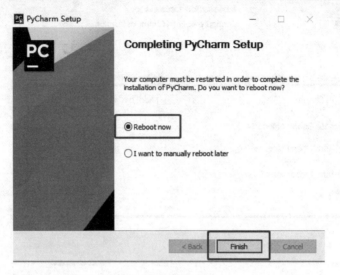

图 1-21　安装完成

步骤 8：重启电脑后，双击桌面上 Pycharm 的快捷方式图标，如图 1-22 所示，选中【Do not import settings】选项，单击【OK】按钮，打开 Pycharm。

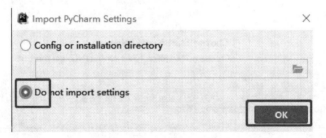

图 1-22　设置导入选项

步骤 9：此时即打开 Pycharm 了，如图 1-23 所示，此 Pycharm 安装完成。

图 1-23　打开 Pycharm

1.3　虚拟环境管理工具 Pipenv 的应用

在 1.2.2 节中已经介绍了同时安装多个 Python 版本的方法了，那么在企业级实战中，如果直接使用安装的 Python 环境是远远不够的，试想一下，在实际代码开发的过程中，必然会遇到这种情况，比如两个项目都使用的是 Python3.9，但是这两个项目对于同一个第三方依赖包的版本要求是不一样的，那么此时如果仅仅使用安装的 Python 环境就无法做到同时开发这两个项目了，因此在实际企业级应用中，一般直接安装 Python 环境都是干净的，在使用的时候，针对每一个项目都要创建一个虚拟环境，这样每个项目之间的第三方依赖包的版本就不会出现冲突的问题了。

而关于虚拟环境的使用在早些时候是通过 virtualenv 来管理的，virtualenv 是通过维护 requirements.txt 文件维护的依赖包的，在实际开发中，特别是涉及跨平台的时候，提交 requirements.txt 管理依赖包的时候总会出现各种各样的问题，在这样的背景下，Pipenv 虚拟环境管理工具应运而生，Pipenv 结合了 Pipfile、Pip、virtualenv，能够有效管理 Python 的多个环境和各种依赖包。

为更好演示 Pipenv 工具的使用，可以先参考 1.2.2 节知识在电脑上安装好 Python3.10、Python3.9、Python3.8、Python3.7 和 Python3.6，因为 Python2.7 官方已经宣布不再支持了，而且当前多数项目也均已经迁移到 Python3.6 及以后的版本了，因此这里就不再演示 Python2.7 的版本了。如图 1-24 所示，此时 Python3.6、Python3.7、Python3.8、Python3.9、Python3.10 均已安装完成。

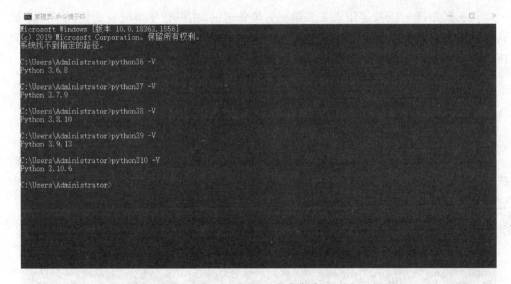

图 1-24　Python 各版本

此时，针对每个 Python 版本的 Pip 工具使用必须加上版本号，如图 1-25 所示。

```
C:\Users\Administrator>pip3.6 -V
pip 18.1 from d:\python36\lib\site-packages\pip (python 3.6)

C:\Users\Administrator>pip3.7 -V
pip 20.1.1 from d:\python37\lib\site-packages\pip (python 3.7)

C:\Users\Administrator>pip3.8 -V
pip 21.1.1 from d:\python38\lib\site-packages\pip (python 3.8)

C:\Users\Administrator>pip3.9 -V
pip 22.0.4 from D:\Python39\lib\site-packages\pip (python 3.9)

C:\Users\Administrator>pip3.10 -V
pip 22.2.1 from D:\Python310\lib\site-packages\pip (python 3.10)

C:\Users\Administrator>
```

图 1-25　Pip 命令

对于使用 Pipenv，不需要每次升级完 Python 版本均使用一次 Pip，大家可以设置一个默认版本的 Pip 命令，比如本书就设置默认的 Pip 命令为 Python3.9 版本的 Pip。当然通过修改环境变量中 Path 的变量也可以，其实还有一个更简单的一个方法，即将其他版本的 Python 安装目录中 Scripts 目录下的 Pip.exe 删除即可，只保留 Python3.9 目录下 Scripts 目录中的 Pip.exe。如图 1-26 所示，此时默认的 Pip 命令指的就是 Python3.9 版本的 Pip 工具了。

```
C:\Users\Administrator>pip -V
pip 22.0.4 from D:\Python39\lib\site-packages\pip (python 3.9)
C:\Users\Administrator>
```

图 1-26　默认的 Pip 命令

众所周知，Pip 工具是用来安装第三方依赖包的，在没有任何配置的情况下，Pip 默认是从官方下载第三方库，下载速度可能会比较慢，因此，一般情况下需要继续配置 Pip 源为国内源，比如阿里云的源。可以使用如下命令配置 Pip 源。

pip config set global.index-url https://mirrors.aliyun.com/pypi/simple/

1.3.1　安装 Pipenv

Pipenv 只需要安装一份即可，因此，在 1.3 节开始的内容中我们已经设置好了默认的 Pip 为 Python3.9 环境下的，这里先使用 Pip list 命令查看当前 Python3.9 环境中已经存在的包，结果如图 1-27 所示，默认情况下只有 Pip 和 setuptools 两个包。

```
C:\Users\Administrator>pip list
Package    Version
---------- -------
pip        22.0.4
setuptools 58.1.0
WARNING: You are using pip version 22.0.4; however, version 22.2.2 is available.
You should consider upgrading via the 'D:\Python39\python.exe -m pip install --upgrade pip' command.
```

图 1-27　执行 Pip list

此外，根据图 1 - 27 中的告警提示，可以执行 Python39 -m pip install -upgrade pip 命令对 Pip 进行升级，如图 1 - 28 所示。

```
C:\Users\Administrator> pip config set global.index-url https://mirrors.aliyun.com/pypi/simple/
Writing to C:\Users\Administrator\AppData\Roaming\pip\pip.ini

C:\Users\Administrator>python39 -m pip install --upgrade pip
Looking in indexes: https://mirrors.aliyun.com/pypi/simple/
Requirement already satisfied: pip in d:\python39\lib\site-packages (22.0.4)
Collecting pip
  Downloading https://mirrors.aliyun.com/pypi/packages/1f/2c/d9626f045e7b49a6225c6b09257861f24da78f4e5f23af2ddbdf852c99b
8/pip-22.2.2-py3-none-any.whl (2.0 MB)
     ---------------------------------------- 2.0/2.0 MB 465.0 kB/s eta 0:00:00
Installing collected packages: pip
  Attempting uninstall: pip
    Found existing installation: pip 22.0.4
    Uninstalling pip-22.0.4:
      Successfully uninstalled pip-22.0.4
Successfully installed pip-22.2.2

C:\Users\Administrator>
```

图 1 - 28　升级 Pip

再次使用 Pip list 命令可以查看到 Pip 的版本已经升级成功，如图 1 - 29 所示。

```
C:\Users\Administrator>pip list
Package    Version
---------- -------
pip        22.2.2
setuptools 58.1.0

C:\Users\Administrator>
```

图 1 - 29　查看 Pip 版本

此时，可以直接使用 Pip install pipenv 进行安装 Pipenv 了，如图 1 - 30 所示，安装完成后再通过 Pip list 可以查看到 Pipenv 已经安装成功了。同时还自动安装了 virtualenv，virtualenv-clone 等其他相关依赖包。

```
C:\Users\Administrator>pip install pipenv
Looking in indexes: https://mirrors.aliyun.com/pypi/simple/
Collecting pipenv
  Downloading https://mirrors.aliyun.com/pypi/packages/8d/4b/a739d5f7cf0a6e2927b2060a99f1793f68f23fc233753ffea90cf414600b/pipenv-2022.8.19-py2.py3-none-any.whl (3.4 MB)
     ---------------------------------------- 3.4/3.4 MB 1.7 MB/s eta 0:00:00
Collecting virtualenv
  Downloading https://mirrors.aliyun.com/pypi/packages/c2/d2/e80b6ad57bc0727fe0bbe7d9c496bde4ef3008e57f452ed1bfafe6618d04/virtualenv-20.16.3-py2.py3-none-any.whl (8.8 MB)
     ---------------------------------------- 8.8/8.8 MB 1.6 MB/s eta 0:00:00
Collecting virtualenv-clone>=0.2.5
  Downloading https://mirrors.aliyun.com/pypi/packages/21/ac/e07058dc5a6c1b97f751d24f20d4b0ec14d735d77f4a1f78c471d6d13a43/virtualenv_clone-0.5.7-py3-none-any.whl (6.6 kB)
Requirement already satisfied: setuptools>=36.2.1 in d:\python39\lib\site-packages (from pipenv) (58.1.0)
Collecting certifi
  Downloading https://mirrors.aliyun.com/pypi/packages/e9/06/d3d367b7af6305b16f0d28ae2aaeb86154fa91f144f036c2d5002a5a202b/certifi-2022.6.15-py3-none-any.whl (160 kB)
     ---------------------------------------- 160.2/160.2 kB 4.8 MB/s eta 0:00:00
Collecting platformdirs<3,>=2.4
  Downloading https://mirrors.aliyun.com/pypi/packages/ed/22/967181c94c3a4063fe64e15331b4cb366bdd7dfbf46fcb8ad89650026fac/platformdirs-2.5.2-py3-none-any.whl (14 kB)
Collecting filelock<4,>=3.4.1
  Downloading https://mirrors.aliyun.com/pypi/packages/94/b3/ff2845971788613e646e667043fdb5f128e2e540aefa09a3c55be8290d6d/filelock-3.8.0-py3-none-any.whl (10 kB)
Collecting distlib<1,>=0.3.5
  Downloading https://mirrors.aliyun.com/pypi/packages/d5/94/db0229ff741c8ff2ff0ac5a8a58d142bd0f81c810701ad9eae2c4c718ede/distlib-0.3.5-py2.py3-none-any.whl (466 kB)
     ---------------------------------------- 467.0/467.0 kB 2.4 MB/s eta 0:00:00
Installing collected packages: distlib, virtualenv-clone, platformdirs, filelock, certifi, virtualenv, pipenv
Successfully installed certifi-2022.6.15 distlib-0.3.5 filelock-3.8.0 pipenv-2022.8.19 platformdirs-2.5.2 virtualenv-20.16.3 virtualenv-clone-0.5.7

C:\Users\Administrator>pip list
Package          Version
---------------- ---------
certifi          2022.6.15
distlib          0.3.5
filelock         3.8.0
pip              22.2.2
pipenv           2022.8.19
platformdirs     2.5.2
setuptools       58.1.0
virtualenv       20.16.3
virtualenv-clone 0.5.7

C:\Users\Administrator>
```

图 1 - 30　安装 Pipenv

1.3.2　Pipenv 工具的使用方法

Pipenv 创建的虚拟环境是针对一个项目的，即针对一个目录，因此使用 Pipenv 创建虚拟环境需要在项目根目录,比如这里在 cmd 窗口进入到 ch/demo -pipenv/ 目录下,然后执行 Pipenv -Python 3.7,即可创建基于 Python37 的虚拟环境(这里需要注意的是此时电脑上必须安装有 Python3.的基本环境)。如图 1 - 31 所示,为自动根据电脑上安装 Python3.6.8 基础环境创建基于 Python3.6.8 的虚拟环境,虚拟环境的位置存放在 C:\Users\Administrator\.virtualenvs\demo-pipenv-u4BrQ_sb 目录下,即创建虚拟环境的目录时以项目名称后加一段随机数组成的文件夹。

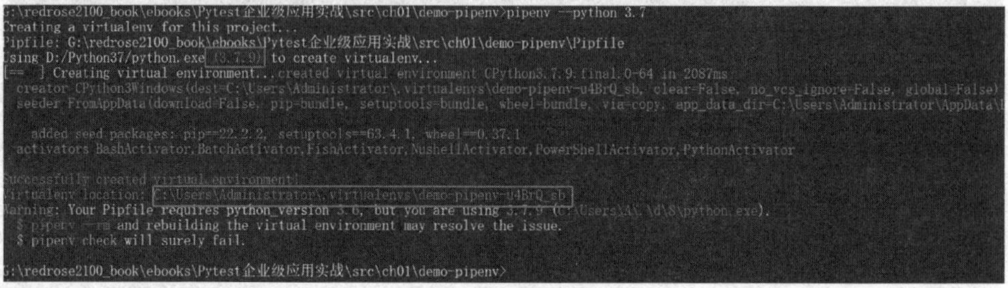

图 1 - 31 创建虚拟环境

通过 Pipenv-venv 命令可以查询虚拟环境的位置,如下所示。

G:\redrose2100_book\ebooks\Pytest 企业级应用实战\src\ch01\demo - pipenv> pipenv -- venv
C:\Users\Administrator\.virtualenvs\demo - pipenv - u4BrQ_sb
G:\redrose2100_book\ebooks\Pytest 企业级应用实战\src\ch01\demo - pipenv>

通过 pipenv run pip list 可以查看当前虚拟环境安装的第三方依赖包,如下所示。

```
G:\redrose2100_book\ebooks\Pytest 企业级应用实战\src\ch01\demo - pipenv> pipenv run pip list
Package      Version
----------   -------
pip          22.2.2
setuptools   63.4.1
wheel        0.37.1


G:\redrose2100_book\ebooks\Pytest 企业级应用实战\src\ch01\demo - pipenv>
```

此时,在项目根目录会自动生成一个 Pipfile,内容如下所示。

```
[[source]]
url = "https://pypi.org/simple"
verify_ssl = true
name = "pypi"


[packages]


[dev - packages]
```

```
[requires]
python_version = "3.7"
```

从这里指定的 Pip 源地址中，可以看到默认给的是官方的地址，为了后续安装第三方库能更快，可以在此文件中将 Pip 源修改为国内的源，比如配置为阿里云的 Pip 源，如下所示。

```
[[source]]
url = "https://mirrors.aliyun.com/pypi/simple/"
verify_ssl = true
name = "pypi"

[packages]

[dev-packages]

[requires]
python_version = "3.7"
```

使用 Pipenv install xxx 即可在此虚拟环境中安装第三方库，比如使用 Pipenv install Pytest 即可在当前虚拟环境中安装 Pytest，如下所示。

```
G:\redrose2100_book\ebooks\Pytest企业级应用实战\src\ch01\demo-pipenv> pipenv install pytest
Installing pytest...
Adding pytest to Pipfile's [packages]...
Installation Succeeded
Pipfile.lock not found, creating...
Locking [packages] dependencies...
Locking...Building requirements...
Resolving dependencies...
Success!
Locking [dev-packages] dependencies...
Updated Pipfile.lock (db8ab1)!
Installing dependencies from Pipfile.lock (db8ab1)...
    ================================0/0 - 00:00:00
To activate this project's virtualenv, run pipenv shell.
Alternatively, run a command inside the virtualenv with pipenv run.

G:\redrose2100_book\ebooks\Pytest企业级应用实战\src\ch01\demo-pipenv>
```

此时，再次通过 Pipenv run pip list 查看当前虚拟环境中已经安装的第三方依赖库，如下所示。此时 Pytest 以及相关的依赖库均已经安装完成了。

```
G:\redrose2100_book\ebooks\Pytest企业级应用实战\src\ch01\demo-pipenv> pipenv run pip list
```

```
Package              Version
-------------------- -------
atomicwrites         1.4.1
attrs                22.1.0
colorama             0.4.5
importlib-metadata   4.12.0
iniconfig            1.1.1
packaging            21.3
pip                  22.2.2
pluggy               1.0.0
py                   1.11.0
pyparsing            3.0.9
pytest               7.1.2
setuptools           63.4.1
tomli                2.0.1
typing_extensions    4.3.0
wheel                0.37.1
zipp                 3.8.1

G:\redrose2100_book\ebooks\Pytest企业级应用实战\src\ch01\demo-pipenv>
```

同时会在该项目的根目录下生成一个Pipfile.lock文件,此文件中记录了当前环境已经安装的第三方库版本号以及源等信息。

至此,Pipenv已经创建了一个虚拟环境,可以看出,虽然此时如果直接使用Pip命令,实际上是Python3.9版本中的Pip,但是通过Pipenv却可以创建一个基于Python3.7的虚拟环境,当然这里需要注意,在电脑上需要事先安装Python3.7,这样以后每个项目中都可以创建一个符合项目要求的Python版本了,而且项目与项目之间的依赖包也是相互独立的,不会出现冲突的问题。

此外,通过Pipenv-rm即可删除当前虚拟环境,这里不再进行删除。我们在前面提到过在使用Pipenv install xxx的时候,会在项目根目录中生成一个Pipfile.lock文件,并且在文件中记录第三方依赖包的版本等信息,这样当项目跨平台时,比如当想在Linux平台上执行时,只需要在参照1.3.1节安装好Pipenv之后,到项目根目录下直接执行Pipenv install命令即可。根据项目根目录下的Pipfile.lock记录的信息可将当前项目所有的第三方依赖库安装上去。这就能实现轻松的跨平台部署。

1.3.3 配置Python解释器

使用Pipenv创建好虚拟环境后,在使用Pycharm进行代码开发时需要首先配置Python解释器,下面是配置解释器的步骤。

步骤1:打开Pycharm,并打开ch01/demo-pipenv项目,依次单击【File】→【Settings】,然后单击【Python Interpreter】→【Add Interpreter】→【Add Local Interpreter】,如图1-32所示。注意,此时虽然Pycharm默认选择了一个解释器,但这并不是我们

想要的,我们希望的是使用通过 Pipenv 创建的虚拟环境。

图 1-32　Settings 配置页面

步骤 2：选中【Virtualenv Environment】,再选中【Existing】选项,然后浏览选择虚拟环境中的 Python.exe 文件,如图 1-33 所示。

图 1-33　设置 Python 解释器

步骤3:如图1-34所示,单击【OK】按钮。

图1-34 选择Python解释器

步骤4:如图1-35所示,此时解释器就设置为Pipenv虚拟环境了,此处也列举出虚拟环境中已经安装的第三方依赖库了,比如Pytest已经在列表中了,单击【OK】按钮,即完成解释器配置。

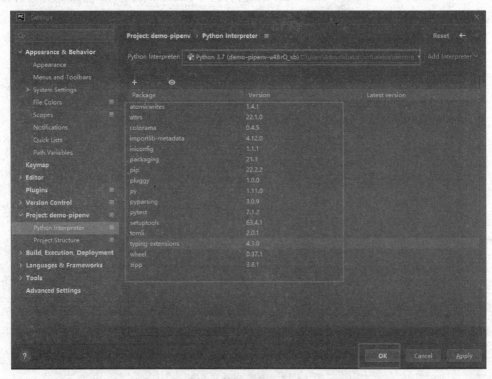

图1-35 虚拟环境中安装的第三方依赖库

步骤 5：此时，在 ch01/demo-pipenv 项目根目录创建 main.py 文件，编写测试代码，如下所示。从代码中调用内置 os 库，查看当前 Python 环境的版本号。

```
import os

if __name__ == "__main__":
    os.system("python -V")
```

然后，用鼠标单击 main.py 文件，执行【run "main"】即可执行 main.py 文件了，执行结果如图 1-36 所示，即当前 Python 环境的版本为 3.7.9。

图 1-36　main.py 执行结果

1.4　Pytest 快速体验

1.4.1　Pytest 安装与升级

（1）安装。

```
pip install pytest
```

（2）查看版本，如下表示安装完成。

```
$ pytest --version
pytest 7.1.2
```

（3）升级。

```
pip install -U pytest
```

1.4.2　创建并执行第一个测试脚本

（1）四行代码即可写一个测试脚本，如下所示，文件命名为 test_demo.py。

```
def func(x):
    return x + 1
```

```
def test_answer():
    assert func(3) == 5
```

（2）直接使用 Pytest 命令即可执行。

```
(demo-HCIhX0Hq) E:\demo> pytest
==================== test session starts ====================
platform win32 -- Python 3.7.9, pytest-7.1.3, pluggy-1.0.0
rootdir: E:\demo
plugins: assume-2.4.3, rerunfailures-10.2
collected 1 item

test_demo.py F                                          [100%]

========================= FAILURES =========================
_____ test_answer _____

    def test_answer():
>       assert func(3) == 5
E       assert 4 == 5
E        +  where 4 = func(3)

test_demo.py:5: AssertionError
================== short test summary info ==================
FAILED test_demo.py::test_answer - assert 4 == 5
==================== 1 failed in 0.06s ====================

(demo-HCIhX0Hq) E:\demo>
```

执行结果中如果 100% 表示测试脚本执行进度，即所有用例均已执行完成，此外，在执行结果最后，Pytest 打印出了执行测试报告，即有一个用例失败，func(3) 的值为 4，与期望值 5 不相等。

这里断言使用了 Assert 语句，关于 Assert 断言语句的高级应用在后续文章中我们会继续深入讲解，这里先体验一下。

1.4.3　Pytest 命令的默认行为

这里 Pytest 命令暂时会执行当前目录以及当前目录所有子目录中的 test_*.py 文件和 *_test.py 文件中的测试脚本，关于更多规则细节我们将在后续篇章中详细展开进行讲解。

1.4.4　对产生的异常进行断言

通过 raise 可以对代码产生的异常进行断言，代码如下所示。

```python
import pytest

def f():
    raise SystemExit(1)

def test_mytest():
    with pytest.raises(SystemExit):
        f()
```

执行 Pytest 命令，结果代码如下所示：

```
(demo-HCIhX0Hq) E:\demo> pytest
==================== test session starts ====================
platform win32 -- Python 3.7.9, pytest-7.1.3, pluggy-1.0.0
rootdir: E:\demo
plugins: assume-2.4.3, rerunfailures-10.2
collected 1 item

test_demo.py .                                        [100%]

===================== 1 passed in 0.01s =====================

(demo-HCIhX0Hq) E:\demo>
```

关于断言的更多的内容，将在后续篇章中详细展开讲解。

1.4.5 在类中编写测试脚本

当编写多个测试脚本时，可以考虑使用类将它们组织起来，这样便于管理。创建测试文件 test_demo.py，并编写类 TestClass 时。注意类名不要以 Test 开头。类中不能有_init__(self)方法，测试类 TestClass2 中有__init__(self)方法，如下所示。

```python
class TestClass:
    def test_one(self):
        x = "this"
        assert "h" in x

    def test_two(self):
        x = "hello"
        assert hasattr(x, "check")

class TestClass2:
    def __init__(self):
        pass
```

```python
    def test_one(self):
        x = "this"
        assert "h" in x

    def test_two(self):
        x = "hello"
        assert hasattr(x, "check")
```

使用 Pytest 命令执行结果如下所示。

```
(demo-HCIhX0Hq) E:\demo> pytest
==================== test session starts ====================
platform win32 -- Python 3.7.9, pytest-7.1.3, pluggy-1.0.0
rootdir: E:\demo
plugins: assume-2.4.3, rerunfailures-10.2
collected 2 items

test_demo.py .F                                         [100%]

========================= FAILURES ==========================
_____ TestClass.test_two _____

self = <test_demo.TestClass object at 0x000001E3C0847848>

    def test_two(self):
        x = "hello"
>       assert hasattr(x, "check")
E       AssertionError: assert False
E        +  where False = hasattr('hello', 'check')

test_demo.py:8: AssertionError
===================== warnings summary ======================
test_demo.py:10
  E:\demo\test_demo.py:10: PytestCollectionWarning: cannot collect test class 'TestClass2' because it has a __init__ constructor (from: test_demo.py)
    class TestClass2:

-- Docs: https://docs.pytest.org/en/stable/how-to/capture-warnings.html
================== short test summary info ==================
FAILED test_demo.py::TestClass::test_two - AssertionError...
========== 1 failed, 1 passed, 1 warning in 0.07s ===========

(demo-HCIhX0Hq) E:\demo>
```

我们发现上述代码中有一个用例通过，还有一个失败，其中还有一个告警，"cannot collect test class 'TestClass2' because it has a __init__ constructor"，即 TestClass2 中无法采集用例，因为它有__init__构造方法。

使用类组织测试函数有许多好处，例如：

可以有效地组织用例；

可以仅在类中共享 fixture；

可以在类上打标签，从而对类中的所有用例打标签。

在使用类组织测试函数时需要特别注意，测试类中的每个用例都是测试类的一个独立的对象，如果类中的用例通过类属性共享变量将是非常糟糕的设计。如下，value 是类变量，会在 test_one 和 test_two 两个测试用例中共享，但是在 test_one 中对 value 进行了修改，通过后续执行 Pytest 命令的时候发现在 test_one 中的修改并未改变 test_two 中的 value 的值，因此，很容易引起误解的。在自动化实践中要尽量避免这类设计。

```python
class TestClassDemoInstance:
    value = 0

    def test_one(self):
        self.value = 1
        assert self.value == 1

    def test_two(self):
        assert self.value == 1
```

执行 Pytest 命令结果如下所示。

```
(demo-HCIhXOHq) E:\demo> pytest
==================== test session starts ====================
platform win32 -- Python 3.7.9, pytest-7.1.3, pluggy-1.0.0
rootdir: E:\demo
plugins: assume-2.4.3, rerunfailures-10.2
collected 2 items

test_demo.py .F                                         [100%]

========================= FAILURES =========================
_____ TestClassDemoInstance.test_two _____

self = <test_demo.TestClassDemoInstance object at 0x000002CD33C1CC08>

    def test_two(self):
>       assert self.value == 1
E       assert 0 == 1
E        +  where 0 = <test_demo.TestClassDemoInstance object at 0x000002CD33C1CC08>.value
```

```
test_demo.py:9: AssertionError
=================== short test summary info ===================
FAILED test_demo.py::TestClassDemoInstance::test_two - as...
================ 1 failed, 1 passed in 0.06s ================
```

(demo-HCIhXOHq) E:\demo>

1.4.6 测试脚本请求,创建一个临时目录

Pytest 提供了许多内置的 fixture,如 tmp_path,将 fixture 名字写入测试函数的参数位置,会在执行测试函数之前率先执行 fixture 的动作,如 tmp_path,即会在执行测试函数之前创建一个临时目录,代码如下所示。

```
def test_needsfiles(tmp_path):
    print(tmp_path)
    assert 0
```

使用 Pytest -q,即输出简要信息执行结果如下所示。

```
(demo-HCIhXOHq) E:\demo> pytest -q
F                                                        [100%]
========================= FAILURES =========================
_____ test_needsfiles _____

tmp_path =
WindowsPath('C:/Users/Administrator/AppData/Local/Temp/pytest-of-Administrator/pytest-18/test_needsfiles0')

    def test_needsfiles(tmp_path):
        print(tmp_path)
>       assert 0
E       assert 0

test_demo.py:3: AssertionError
-------------------- Captured stdout call --------------------
C:\Users\Administrator\AppData\Local\Temp\pytest-of-Administrator\pytest-18\test_needsfiles0
=================== short test summary info ===================
FAILED test_demo.py::test_needsfiles - assert 0
1 failed in 0.06s
```

(demo-HCIhXOHq) E:\demo>

由上述我们可以看到在执行测试函数之前创建了临时目录。

这里我们仅演示 fixture 的神奇的作用,在后续篇章中将对 fixture 更多的功能详细展开讲解。

第 2 章 Pytest 脚本规则、组成与运行

2.1 Pytest 脚本的规则

2.1.1 测试脚本文件命名规则

如下所示,测试脚本文件为 Python 文件,此外,文件命名规则为 test_*.py 或者 *_test.py 格式的文件,均为符合 Pytest 要求的测试文件命名规范:

test_demo.py
test_01.py
test_.py
demo_test.py
01_test.py
_test.py

如果不符合 Pytest 要求的文件命名规则,Pytest 会找不到测试脚本如下所示。

test.py
testdemo.py
Test_demo.py
TestDemo.py
Test_.py
Demo_Test.py
_Test.py
Test.py
Demo.py

这里面有一点需要格外注意,由于 Windows 上文件名是不区分大小写的,默认情况下 Windows 会把所有字符转换为小写处理,因此如果在 Windows 上调试,可能会发现类似的 Test_demo.py 文件 Pytest 是有用例的,有这种情况是 Windows 的特殊处理导致的,因此这里一定不要混淆,否则当后续将脚本迁移到 linux 服务器上执行的时候就会出现比如 Windows 上能执行在 linux 上不执行的异常事件。这里统一按照标准处理,即测试文件必须为 test_*.py 或者 *_test.py 格式。

2.1.2 测试函数的测试类命名规则

在测试脚本中,测试函数分为两类,一种是直接定义在测试文件中的,比如:

```
def test_func():
    assert 1 == 1
```

另一种则是使用类组织的在类内的测试函数,如下所示。

```
class TestDemo:
    def test_func():
        assert 1 == 1
```

为了方便记忆,不论类外还是类内统一称为测试函数,如此方便将测试类和测试函数命名规则总结为:

(1) 测试函数名必须以 test 开头。
(2) 测试类名必须以 test 开头。
(3) 测试类中不能有 __init__(self) 方法。

如下的测试函数均为符合 Pytest 规则的测试函数。

```
def test_demo():
    assert 1 == 1

def testdemo():
    assert 1 == 1

def test():
    assert 1 == 1

def test_():
    assert 1 == 1
```

如果测试函数均为不符合 Pytest 规则的函数,就不会被 Pytest 发现。

```
def demo_test():
    assert 1 == 1

def demotest():
    assert 1 == 1

def _test():
    assert 1 == 1

def Test():
    assert 1 == 1
```

```
def Test_Demo():
    assert 1 == 1

def TestDemo():
    assert 1 == 1

def DemoTest():
    assert 1 == 1
```

对于使用类组织的测试函数，类必须首先满足要求，即类型以 Test 开头，并且类中没有 __init__ 方法，类中的测试函数名再符合测试函数的命名规则，即以 test 开头时，才会被认为是测试脚本，如下所示。

```
class TestDemo:
    def test_demo(self):
        assert 1 == 1

class Test:
    def test_demo(self):
        assert 1 == 1
```

如果类名不是以 Test 开头或者类名以 test 开头，但是类中有 __init__ 方法时，不论类中的测试函数名是否符合 Pytest 要求的规则，均不会被 Pytest 识别。

下面以一个综合的例子来演示一下，我们为了演示具体哪个测试函数能被 Pytest 识别，在每个测试函数中加了打印信息，如下所示。

```
def demo_test():
    print("in demo_test")
    assert 1 == 1

def test_demo():
    print("in test_demo")
    assert 1 == 1

class TestDemo:
    def __init__(self):
        pass

    def test_demo(self):
        print("in TestDemo.test_demo")
        assert 1 == 1
```

```
class Test:
    def test_demo(self):
        print("in Test.test_demo")
        assert 1 == 1
```

使用 Pytest -s 命令执行结果如下。加 -s 参数是为了将打印信息打印出来。

```
$ pytest -s
=========================== test session starts ===========================
platform win32 -- Python 3.9.13, pytest-7.1.2, pluggy-1.0.0
rootdir: D:\redrose2100-book\ebooks\Pytest 企业级应用实战\src
collected 2 items

test_demo.py in test_demo
.in Test.test_demo
.

============================ warnings summary ============================
test_demo.py:10
  D:\redrose2100-book\ebooks\Pytest 企业级应用实战\src\test_demo.py:10: PytestCollectionWarning: cannot collect test class 'TestDemo' because it has a __init__ constructor (from: test_demo.py)
    class TestDemo:

-- Docs: https://docs.pytest.org/en/stable/how-to/capture-warnings.html
======================= 2 passed, 1 warning in 0.02s =======================
```

可以看出，这里打印出了"in test_demo"和"in Test.test_demo"，只有这两个测试函数被 Pytest 识别了，此外，还打印出了一条告警信息"cannot collect test class 'TestDemo' because it has a __init__ constructor"，即提示 TestDemo 类中因为有 __init__ 构造函数，因而无法识别其中的测试函数。

2.1.3 测试脚本目录的约束条件

Pytest 对测试脚本目录的文件夹命名没有什么要求，但是有如下两条约定：
- 如果每一个文件夹中均创建 __init__.py 文件，则没有其他要求。
- 如果每一个文件夹中没有创建 __init__.py 文件，此时，所有的测试脚本文件名必须独一无二，否则 Pytest 会报错。

首先，按照如下结构创建文件及文件夹目录。

```
base
 |--------demo01
            |--------__init__.py
            |--------test_demo.py
```

```
        |--------demo02
                |--------__init__.py
                |--------test_demo.py
```

在每个test_demo.py文件中编写一条最简单的测试脚本,如下所示。

```
def test_demo():
    assert 1 == 1
```

执行Pytest命令结果如下:

```
$ pytest
========================== test session starts ==========================
platform win32 -- Python 3.9.13, pytest-7.1.2, pluggy-1.0.0
rootdir: D:\redrose2100-book\ebooks\Pytest企业级应用实战\src
collected 2 items

demo01\test_demo.py .                                              [ 50%]
demo02\test_demo.py .                                              [100%]

=========================== 2 passed in 0.11s ===========================
```

此时,虽然两个测试文件的文件名相同,但因为demo01和demo02文件夹中均有__init__.py文件,因此,此时执行Pytest是不会报错的,即当文件夹中均有__init__.py文件时,就没有其他要求或约束。

下面将demo01和demo02文件夹中的__init__.py文件均删除,目录结构如下所示。

```
base
    |--------demo01
            |--------test_demo.py
    |--------demo02
            |--------test_demo.py
```

此时执行Pytest命令结果如下所示。

```
$ pytest
_____ ERROR collecting demo02/test_demo.py _____
import file mismatch:
imported module 'test_demo' has this __file__ attribute:
    D:\redrose2100-book\ebooks\Pytest企业级应用实战\src\demo01\test_demo.py
which is not the same as the test file we want to collect:
    D:\redrose2100-book\ebooks\Pytest企业级应用实战\src\demo02\test_demo.py
HINT: remove __pycache__ / .pyc files and/or use a unique basename for your test file modules
========================= short test summary info =========================
ERROR demo02/test_demo.py
```

```
!!!!!!!!!!!!!!!!!!!!!!!!!! Interrupted: 1 error during collection !!!!!!!!!!!!!!!!!!!!!!!!!!
========================== 1 error in 0.10s ==============================
```

可以发现，此时出现报错提示，提示文件名要用独一无二的名称。这因为是跟Pytest加载用例的原理有关。Pytest在加载用例的时候实质上是把测试用例文件当作模块导入的，而在导入模块之前，Pytest先把一个basedir的变量添加到环境变量中，basedir会针对每个测试脚本，向上递归查找，找到最顶层带有__init__.py文件的目录，然后这个目录的上层目录便作为basedir添加到环境变量中，之后根据basedir计算每个测试脚本文件的相对路径。这么说比较枯燥，我们以下面的目录结构为例说明。

我们针对第一个test_demo.py文件，从当前目录往上找，会找到最后一层带有__init__.py文件的目录，由于demo01本身就没有__init__.py文件，因此Pytest会把demo01的目录添加到环境变量中，然后从demo01开始确定第一个test_demo.py导入的相对路径，即test_demo.py。

对于第二个test_demo.py文件，同样从当前目录开始往上找，找到最后一层带有__init__.py文件的目录，由于demo02本身也没有__init__.py文件，此时Pytest会把这个目录demo02添加到环境变量中，然后从demo02开始计算第二个test_demo.py导入的相对路径，即test_demo.py，此时这两个相同名字的文件虽然在不同层次不同文件夹中，但是经过Pytest计算，我们会发现这两个导入路径完全相同，因此此时会报错。

至于Pytest为什么会这么计算，我们将在后续Pytest脚本加载原理中进行详细的讲解，这里暂时不再深入探讨了。

2.1.4 自定义测试文件名、测试类、测试函数命名规则

一般情况下，通过Pytest.ini可以自定义测试文件、测试类、测试函数的命名规则，不建议去修改测试文件名、测试类、测试函数默认的命名规则，因为Pytest定义的默认规则，相当于定义好了协议，大家都是认可和熟知的，假如重新定义了命名规则，则必然带来一个问题，即其他人不了解，也不清楚当前的规则是什么样的，这样就会带来了许多不必要的麻烦。

在有些较大的公司，他们可能会根据公司业务的特点，制定公司级的标准，同时对默认的命名规则进行重新定义。定义的方法很简单，比如某公司觉得测试文件名和测试函数名的规则不一致很容易导致混乱，因此想根据自己的需求统一一下，即约束测试文件命名规则为check_*.py格式，测试类以Check开头，测试函数为check_*格式，这样就做到了形式上的统一。此时，只需要在脚本的根目录下的Pytest.ini文件中增加如下配置即可。

```
[pytest]
python_files = check_*.py
python_classes = Check
python_functions = check_*
```

下面就以一个实例演示文件目录的结构。

```
base
  |----pytest.ini
  |----check_demo.py
```

其中 Pytest.ini 的配置代码为：

```
[pytest]
python_files = check_*.py
python_classes = Check
python_functions = check_*
```

check_demo.py 中的测试函数和测试类的代码如下所示。

```
def check_func():
    assert 1 == 1


class CheckDemo():
    def check_func2(self):
        assert 1 == 1
```

从执行结果可以看出此时测试文件已经不是默认命名规则了，测试类和测试函数的命名均已不是 Pytest 默认的命名规则，而是自定义的规则，而 Pytest 此时依然可以识别，并且执行，这就是 Pytest.ini 配置的原因。

```
$ pytest
=========================== test session starts ===========================
platform win32 -- Python 3.9.13, pytest-7.1.2, pluggy-1.0.0
rootdir: G:\redrose2100_book\ebooks\Pytest 企业级应用实战\src, configfile: pytest.ini
collected 2 items

check_demo.py ..                                                    [100%]

============================ 2 passed in 0.01s ============================
```

2.2　Pytest 脚本的组成

2.2.1　自动化脚本组成简介

一般来说，自动化脚本是由若干个测试套组成的，而每个测试套又由测试套 setup 和若干个脚本，以及测试套 teardown 组成的。每个脚本又是由测试脚本级 setup 和测试脚本正文、测试脚本级 teardown 组成，具体结构如下所示。

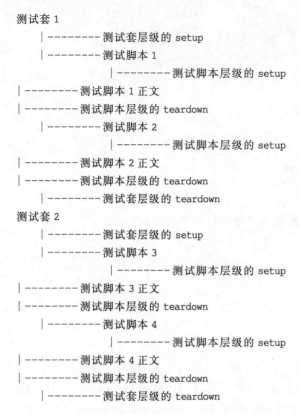

在 Pytest 自动化测试框架中，测试套可以是一个类，也可以是一个文件，测试脚本可以是类中的一个测试函数，也可以是文件中的一个测试函数，下面就将 Pytest 中的测试函数、测试类以及各个层级的 setup 和 teardown 详细展开讲解。

2.2.2 测试类中各个层级的 setup 和 teardown

在 Pytest 中，如果使用类组织测试脚本，则一个类相当于一个测试套，这样的测试套有测试套级别的 setup 和 teardown。而类级别的 setup 和 teardown 的名字分别为 setup_class 和 teardown_class，因为类中的函数准确地说应该叫方法，所以类中的测试脚本级的 setup 和 teardown 的名字分别为 setup_method 和 teardown_method。如果代码中有两个测试函数 test_01 和 test_02，因此，从理论上分析，代码的执行顺序应首先执行测试套级的 setup_class。之后执行第一个脚本 setup_method, test_01, teardown_method。接下来执行第二个脚本，即 setup_method, test_02, teardown_method，最后，执行测试套级的 teardown_class。

```
class TestDemo:
    def setup_class(self):
        print("in setup_class")

    def teardown_class(self):
```

```python
        print("in testdown_class")

    def setup_method(self):
        print("in setup_method")

    def teardown_method(self):
        print("in teardown_method")

    def test_01(self):
        print("in test_01")
        assert 1 == 1

    def test_02(self):
        print("in test_02")
```

使用 Pytest -s 命令执行结果如下面代码所示,通过打印的信息可以看出执行顺序与前面理论分析的顺序是一致的。

```
$ pytest -s
=========================== test session starts ===========================
platform win32 -- Python 3.9.13, pytest-7.1.2, pluggy-1.0.0
rootdir: D:\redrose2100-book\ebooks\Pytest 企业级应用实战\src
collected 2 items

test_demo.py in setup_class
in setup_method
in test_01
.in teardown_method
in setup_method
in test_02
.in teardown_method
in testdown_class


============================ 2 passed in 2.14s ===========================
```

2.2.3　测试文件中各个层级的 setup 和 teardown

测试文件又叫测试模块,即 module。在 Pytest 框架中,一个测试文件也可以作为一个测试套来组织管理测试脚本,如果代码中一个测试文件组织管理了两个测试脚本 test_03 和 test_04,既然是测试套就有测试套层级 setup 和 teardown 的概念,在这里,测试套层级的 setup 和 teardown 的名字为 setup_module 和 teardown_module,在 Python 文件中的函数叫作函数,所以测试模块中测试脚本层级的 setup 和 teardown 的名字为 setup_function 和 teardown_function。如下面代码所示，从理论上分析，应首先执行测试套的 setup，即 setup_module。然后执行第一个测试脚本，即 setup_function，test_03，teardown_function，再执行第二个脚本，即 setup_function，test_04，teardown_function。最后执行测试套级别的 teardown，即 teardown_module。

```python
def setup_module():
    print("in setup_module")

def teardown_module():
    print("in teardown_module")

def setup_function():
    print("in setup_function")

def teardown_function():
    print("in teardown_function")

def test_03():
    print("in test_03")
    assert 1 == 1

def test_04():
    print("in test_04")
    assert 1 == 1
```

使用 Pytest -s 命令执行结果如下，可以看出，通过打印的顺序这里的执行顺序与前面的理论分析的顺序是完全一致的。

```
$ pytest -s
============================ test session starts ============================
platform win32 -- Python 3.9.13, pytest-7.1.2, pluggy-1.0.0
rootdir: D:\redrose2100-book\ebooks\Pytest 企业级应用实战\src
collected 2 items

test_demo.py in setup_module
in setup_function
in test_03
.in teardown_function
in setup_function
in test_04
.in teardown_function
in teardown_module


============================ 2 passed in 2.07s ============================
```

2.2.4 测试文件中测试函数和类中测试方法混合时各个层级的 setup 与 teardown

当测试文件中测试函数和类中的测试方法混合时，执行顺序看上去会非常复杂。其实这里面可以把测试类 TestDemo 看作与 test_03 和 test_04 同一级别进行分析，这样就很容分析出执行的顺序了，首先执行测试模块级的 setup，即 setup_module，然后执行第一个脚本，其中，第一个可以理解为 TestDemo 测试，此时就可以按照测试类的执行原理将 TestDemo 中的代码执行完毕。再执行第二个脚本，即 setup_function，test_03，teardown_function。接下来执行第三个脚本，即 setup_function，test_04，teardown_function，最后执行测试模块级的 teardown_module。

```python
class TestDemo:
    def setup_class(self):
        print("in setup_class")

    def teardown_class(self):
        print("in testdown_class")

    def setup_method(self):
        print("in setup_method")

    def teardown_method(self):
        print("in teardown_method")

    def test_01(self):
        print("in test_01")
        assert 1 == 1

    def test_02(self):
        print("in test_02")

def setup_module():
    print("in setup_module")

def teardown_module():
    print("in teardown_module")

def setup_function():
    print("in setup_function")

def teardown_function():
    print("in teardown_function")
```

```
def test_03():
    print("in test_03")
    assert 1 == 1

def test_04():
    print("in test_04")
    assert 1 == 1
```

使用 Pytest-s 命令执行结果如下所示,通过打印内容可以看出执行顺序与前面的理论分析的顺序是一致的。

```
$ pytest -s
=========================== test session starts ===========================
platform win32 -- Python 3.9.13, pytest-7.1.2, pluggy-1.0.0
rootdir: D:\redrose2100-book\ebooks\Pytest 企业级应用实战\src
collected 4 items

test_demo.py in setup_module
in setup_class
in setup_method
in test_01
.in teardown_method
in setup_method
in test_02
.in teardown_method
in testdown_class
in setup_function
in test_03
.in teardown_function
in setup_function
in test_04
.in teardown_function
in teardown_module


============================ 4 passed in 0.02s ============================
```

2.2.5 各个层级的 setup 和 teardown 在自动化实践中的应用

通过前面对各个层级 setup 和 teardown 执行顺序的分析,可以看出,测试套层级的 setup 主要用于测试套中所有测试脚本公共而且只需要一次的配置,而测试套层级的 teardown 则用于测试套内所有用例执行完成后恢复环境或者清理的配置。比如当一个测试套中的所有用例都需要某个用户登录之后才能做时,登录操作就可以在测试套级别的 setup 中实现,退出登录操作则可在测试套级别的 teardown 中实现。而对于测试套中的每个用例,当用户登录后,都希望在每个用例中首先将页面切换到首页,然后再进行相关的操作,此时切换到首页的动作就可以在测试脚本级的 setup 中实现,即只需要实现一次。当前测试套中的每个用例执行之前都会首先执行脚本级 setup 操作,即切换到首页,这样就减少了测试脚本的代码重复率,同时也提高了脚本的运行效率。

当然,在自动化实践应用中,需要提前给测试人员传达上述思想,目的提醒大家测试在设计用例的时候要把用例按照测试套的维度划分。

2.3 Pytest 脚本的运行

2.3.1 指定目录或文件

Pytest 命令后面可以指定文件或者目录,当指定文件时,Pytest 会执行指定文件中的所有符合 Pytest 规则的测试函数或测试方法。当指定目录时,则会递归地从指定目录中寻找符合条件的测试文件,然后执行测试文件中的所有测试函数或测试方法。下面我们目录结构介绍 Pytest 命令如何指定文件或目录执行测试脚本。

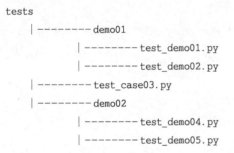

当使用 Pytest test_case03.py 命令时,即指定测试文件,此时,会执行 test_case.03.py 文件中的所有测试函数或测试方法。当使用 Pytest demo01 命令时,即指定目录,此时会执行 demo01 目录下的所有测试文件中的测试函数和测试方法,而当直接执行 Pytest 命令时,相当于指定了当前目录,即此时会执行当前目录中以及所有子目录中的所有测试文件的测试函数或者测试方法。

2.3.2 指定测试函数或测试方法

Pytest 命令也可以直接指定具体的测试函数或测试方法以及测试类,指定测试函数时只执行指定的一个测试函数。如果指定类中的测试方法也执行指定的一个测试方法,那么当指定测试类时,则执行指定类中的所有测试方法。下面以测试代码演示如何指定测试函数或测试方法，test_demo01.py 代码如下所示。

```python
def test_func1():
    print("\nin test_demo01.test_func1......")
    assert 1 == 1

def test_func2():
    print("\nin test_demo01.test_func2......")
    assert 1 == 1

class TestDemo(object):
    def test_func1(self):
        print("\nin test_demo01.TestDemo.test_func1......")
        assert 1 == 1

    def test_func2(self):
        print("\nin test_demo01.TestDemo.test_func2......")
        assert 1 == 1
```

如下指定测试函数时只执行指定的 test_func1 函数。

```
$ pytest -s test_demo01.py::test_func1
============================ test session starts =============================
platform win32 -- Python 3.9.7, pytest-6.2.5, py-1.10.0, pluggy-1.0.0
rootdir: D:\src\blog\tests\demo01
plugins: allure-pytest-2.9.43, caterpillar-pytest-0.0.2, forked-1.3.0, rerunfailures-10.1, xdist-2.3.0
collected 1 item

test_demo01.py
in test_demo01.test_func1......
.

============================ 1 passed in 0.02s =============================
$
```

指定测试类中的测试方法时，只会执行指定的 test_func1 方法。

```
$ pytest -s test_demo01.py::TestDemo::test_func1
============================ test session starts =============================
platform win32 -- Python 3.9.7, pytest-6.2.5, py-1.10.0, pluggy-1.0.0
rootdir: D:\src\blog\tests\demo01
plugins: allure-pytest-2.9.43, caterpillar-pytest-0.0.2, forked-1.3.0, rerunfailures-10.1, xdist-2.3.0
collected 1 item
```

```
test_demo01.py
in test_demo01.TestDemo.test_func1......
.

=============================1 passed in 0.01s ============================

$
```

当指定测试类时,则会执行整个类中的所有测试方法。

```
$ pytest -s test_demo01.py::TestDemo
============================= test session starts =========================
platform win32 -- Python 3.9.7, pytest-6.2.5, py-1.10.0, pluggy-1.0.0
rootdir: D:\src\blog\tests\demo01
plugins: allure-pytest-2.9.43, caterpillar-pytest-0.0.2, forked-1.3.0, rerunfail-
ures-10.1, xdist-2.3.0
collected 2 items

test_demo01.py
in test_demo01.TestDemo.test_func1......

in test_demo01.TestDemo.test_func2......
.

=============================2 passed in 0.02s ============================

$
```

2.3.3 通过--k参数对文件类名及函数名进行模糊匹配和挑选

Pytest-k 挑选脚本执行是非常灵活的,Pytest-k 后面跟的是一个字符串,会从当前路径递归到子目录寻找测试脚本。当给定的字符串匹配一个目录时,则此目录下所有用例均会被执行。当给定的字符串未在目录名匹配中匹配,则会继续找目录中的测试文件。当测试文件匹配后,则此测试文件中的所有测试函数或测试方法均会被执行,而如果测试文件名未匹配时,则继续匹配测试文件中的测试类、测试函数、测试方法。

比如有目录结构如下所示。

```
demo01------
    |---------test_demo01.py
    |---------test_demo02.py
!!!
```

其中 test_demo01.py 内容如下:

```
def test_func1():
    print("\nin test_demo01.test_func1......")
    assert 1 == 1

def test_func2():
    print("\nin test_demo01.test_func2......")
    assert 1 == 1

class TestDemo(object):
    def test_func1(self):
        print("\nin test_demo01.TestDemo.test_func1......")
        assert 1 == 1

    def test_func2(self):
        print("\nin test_demo01.TestDemo.test_func2......")
        assert 1 == 1
```

test_demo02.py 内容如下:

```
def test_func1():
    print("\nin test_demo02.test_func1......")
    assert 1 == 1

def test_func2():
    print("\nin test_demo02.test_func2......")
    assert 1 == 1
```

当执行如下命令时,因为目录 demo01 中已经存在"1"了,此时,demo01 内的所有用例均会被选中执行。

```
$ pytest -s -k "1"
============================ test session starts ============================
platform win32 -- Python 3.9.7, pytest-6.2.5, py-1.10.0, pluggy-1.0.0
rootdir: D:\src\blog\tests\demo01
plugins: allure-pytest-2.9.43, caterpillar-pytest-0.0.2, forked-1.3.0, rerunfailures-10.1, xdist-2.3.0
collected 6 items / 1 deselected / 5 selected

test_demo01.py
in test_demo01.test_func1......
.
in test_demo01.test_func2......
.
in test_demo01.TestDemo.test_func1......
```

```
in test_demo01.TestDemo.test_func2......

test_demo02.py
in test_demo02.test_func1......
.
======================= 5 passed, 1 deselected in 0.02s =======================
$
```

当执行如下命令时，发现 demo02 的目录名中没有 2，因此继续找 demo01 中的文件，显然 test_demo01.py 文件名中也没有 2，则此时继续进入 test_demo01.py。如果测试函数 test_func2 命中，即会被执行，而 TestDemo 类型没有匹配，则会继续在 TestDemo 类中找，此时 TestDemo 类中的 test_func2 测试方法命中，即在 test_demo01.py 文件中命中了一个测试函数和一个测试方法。当 test_demo02.py 文件名本身已经含有 2 了，则 test_demo02.py 中的所有测试函数和测试方法均会被执行，执行结果与上述分析一致。

```
$ pytest -s -k "2"

demo01\test_demo01.py
in test_demo01.test_func2......
.
in test_demo01.TestDemo.test_func2......
.
demo01\test_demo02.py
in test_demo02.test_func1......
.
in test_demo02.test_func2......
.
======================= 4 passed, 2 deselected in 2.22s =======================
$
```

此外，Pytest -k 后面的表达式还支持逻辑关系匹配，注意是匹配 1 但是不匹配 func2，执行结果如下所示。

```
$ pytest -s -k "1 and not func2"
=========================== test session starts ===========================
platform win32 -- Python 3.9.7, pytest-6.2.5, py-1.10.0, pluggy-1.0.0
rootdir: D:\src\blog\tests\demo01
plugins: allure-pytest-2.9.43, caterpillar-pytest-0.0.2, forked-1.3.0, rerunfailures-10.1, xdist-2.3.0
```

```
collected 6 items / 3 deselected / 3 selected

test_demo01.py
in test_demo01.test_func1......
.
in test_demo01.TestDemo.test_func1......
.
test_demo02.py
in test_demo02.test_func1......


======================= 3 passed, 3 deselected in 0.02s =====================
$
```

2.3.4 通过--ignore参数挑选用例,忽略执行

--ignore参数相当于挑选出一部分用例不执行,比如--ignore给定的是目录,则此目录下的所有测试函数和测试方法均不会执行,如果--ignore后面给定的是测试文件,则此测试文件中的所有测试方法和测试函数均不会执行。

目录结构如下所示。

```
tests/
|-- example
|   |-- test_example_01.py
|   |-- test_example_02.py
|   '-- test_example_03.py
|-- foobar
|   |-- test_foobar_01.py
|   |-- test_foobar_02.py
|   '-- test_foobar_03.py
'-- hello
    '-- world
        |-- test_world_01.py
        |-- test_world_02.py
        '-- test_world_03.py
```

使用Pytest-ignore=tests/foobar/test_foobar_03.py--ignore=tests/hello/时,可以用多个ignore参数去忽略,我们可以看到tests/foobar/test_foobar_03.py和hello目录下的参数都没有执行,如下所示。

```
$ pytest --ignore=foobar/test_foobar_03.py --ignore=hello/
=========================== test session starts ===========================
platform win32 -- Python 3.9.6, pytest-6.2.5, py-1.10.0, pluggy-1.0.0
```

```
    rootdir: G:\src\blog\tests
    plugins: allure-pytest-2.9.43, caterpillar-pytest-0.0.2, forked-1.3.0, rerunfail-
ures-10.1, xdist-2.3.0
    collected 5 items

    example\test_example_01.py .                                              [ 20%]
    example\test_example_02.py .                                              [ 40%]
    example\test_example_03.py .                                              [ 60%]
    foobar\test_foobar_01.py .                                                [ 80%]
    foobar\test_foobar_02.py .                                                [100%]

    ============================5 passed in 0.05s ============================

$
```

2.3.5 通过--ignore-glob参数支持正则挑选用例忽略

--ignore-glob 参数的作用与--ignore 的作用相同,都是忽略挑选出来的用例,不同的是,--ignore 参数需要指定的是目录或文件或测试函数,而--ignore-glob 参数则可以通过正则的方式进行匹配,对匹配上的脚本均忽略不执行。

目录结构如下所示。

```
tests/
|-- example
|   |-- test_example_01.py
|   |-- test_example_02.py
|   '-- test_example_03.py
|-- foobar
|   |-- test_foobar_01.py
|   |-- test_foobar_02.py
|   '-- test_foobar_03.py
'-- hello
    '-- world
        |-- test_world_01.py
        |-- test_world_02.py
        '-- test_world_03.py
```

在 Unix shell 风格下,使用--ignore-glob 参数,可以通过正则匹配的方式忽略用例如下所示,以"_01.py"结尾的用例均已被忽略。

```
$ pytest --ignore-glob='*_01.py'
============================ test session starts =========================
platform win32 -- Python 3.9.6, pytest-6.2.5, py-1.10.0, pluggy-1.0.0
rootdir: G:\src\blog\tests
```

```
plugins: allure-pytest-2.9.43, caterpillar-pytest-0.0.2, forked-1.3.0, rerunfail-
ures-10.1, xdist-2.3.0
collected 6 items

example\test_example_02.py .                                          [ 16%]
example\test_example_03.py .                                          [ 33%]
foobar\test_foobar_02.py .                                            [ 50%]
foobar\test_foobar_03.py .                                            [ 66%]
hello\workd\test_world_02.py .                                        [ 83%]
hello\workd\test_world_03.py .                                        [100%]

============================ 6 passed in 0.06s ============================
```

2.3.6　通过--deselect参数挑选用例不执行,并显示未执行数量

--deselect参数跟--ignore参数作用类似,都是挑选一部分用例不执行,不同的是,--ignore参数执行完成后只显示执行了多少条用例,而--deselect会显示有多少条用例被选中未执行,如下所示。

```
tests/
|-- example
|   |-- test_example_01.py
|   |-- test_example_02.py
|   '-- test_example_03.py
|-- foobar
|   |-- test_foobar_01.py
|   |-- test_foobar_02.py
|   '-- test_foobar_03.py
'-- hello
    '-- world
        |-- test_world_01.py
        |-- test_world_02.py
        '-- test_world_03.py
```

有些命令可不执行test_foobar_01.py和test_foobar_03.py文件的用例,如果忽略多个命令可以使用多个--deselect参数。

```
$ pytest --deselect foobar/test_foobar_01.py --deselect foobar/test_foobar_03.py
============================ test session starts ============================
platform win32 -- Python 3.9.6, pytest-6.2.5, py-1.10.0, pluggy-1.0.0
rootdir: G:\src\blog\tests
plugins: allure-pytest-2.9.43, caterpillar-pytest-0.0.2, forked-1.3.0, rerunfail-
ures-10.1, xdist-2.3.0
collected 9 items / 2 deselected / 7 selected
```

```
example\test_example_01.py .                                    [ 14%]
example\test_example_02.py .                                    [ 28%]
example\test_example_03.py .                                    [ 42%]
foobar\test_foobar_02.py .                                      [ 57%]
hello\workd\test_world_01.py .                                  [ 71%]
hello\workd\test_world_02.py .                                  [ 85%]
hello\workd\test_world_03.py .                                  [100%]

====================== 7 passed, 2 deselected in 0.07s ======================
```

2.3.7 通过重复指定文件路径,实现用例重复执行

通过重复指定文件路径可实现用例重复执行,目录结构如下。

```
tests/
|-- foobar
|    |-- test_foobar_01.py
|    |-- test_foobar_02.py
|    '-- test_foobar_03.py
```

当指定的重复路径为文件级别时,将默认支持执行多次,执行了两次的情况如下所示。

```
$ pytest  foobar/test_foobar_01.py foobar/test_foobar_01.py
============================ test session starts ============================
platform win32 -- Python 3.9.6, pytest-6.2.5, py-1.10.0, pluggy-1.0.0
rootdir: G:\src\blog\tests
plugins: allure-pytest-2.9.43, caterpillar-pytest-0.0.2, forked-1.3.0, rerunfailures-10.1, xdist-2.3.0
collected 2 items


foobar\test_foobar_01.py ..


============================ 2 passed in 0.02s ============================
```

指定的重复路径为目录时,默认只会执行一次,只执行了一次的情况如下所示。

```
$ pytest  foobar foobar
============================ test session starts ============================
platform win32 -- Python 3.9.6, pytest-6.2.5, py-1.10.0, pluggy-1.0.0
rootdir: G:\src\blog\tests
plugins: allure-pytest-2.9.43, caterpillar-pytest-0.0.2, forked-1.3.0, rerunfailures-10.1, xdist-2.3.0
collected 3 items
```

```
foobar\test_foobar_01.py .                                          [ 33%]
foobar\test_foobar_02.py .                                          [ 66%]
foobar\test_foobar_03.py .                                          [100%]

============================ 3 passed in 0.03s ============================
```

当指定的路径为目录时,如果希望也执行多次,需要使用--keep-duplicates参数,如目录中的用例执行了两次的情况如下所示。

```
$ pytest  foobar foobar -- keep-duplicates
============================ test session starts ============================
platform win32 -- Python 3.9.6, pytest-6.2.5, py-1.10.0, pluggy-1.0.0
rootdir: G:\src\blog\tests
plugins: allure-pytest-2.9.43, caterpillar-pytest-0.0.2, forked-1.3.0, rerunfailures-10.1, xdist-2.3.0
collected 6 items

foobar\test_foobar_01.py .                                          [ 16%]
foobar\test_foobar_02.py .                                          [ 33%]
foobar\test_foobar_03.py .                                          [ 50%]
foobar\test_foobar_01.py .                                          [ 50%]
foobar\test_foobar_02.py .                                          [ 50%]
foobar\test_foobar_03.py .

============================ 6 passed in 0.04s ============================
```

2.3.8　通过--collect-only参数不执行脚本而统计脚本总数

当需要查看有多少用例,但又不想通过执行所有用例时,可以使用--collect-only参数,目录结构如下所示。

```
tests/
|-- pytest.ini
|-- example
|   |-- test_example_01.py
|   |-- test_example_02.py
|   '-- test_example_03.py
|-- foobar
|   |-- test_foobar_01.py
|   |-- test_foobar_02.py
|   '-- test_foobar_03.py
'-- hello
    '-- world
```

```
|-- test_world_01.py
|-- test_world_02.py
'-- test_world_03.py
```

执行 Pytest -- collect-only 查看当前目录下有多少用例。

```
$ pytest -- collect-only
============================== test session starts =======================
platform win32 -- Python 3.9.6, pytest-6.2.5, py-1.10.0, pluggy-1.0.0 -- D:\python39\python.exe
cachedir: .pytest_cache
hypothesis profile 'default' ->
database = DirectoryBasedExampleDatabase('G:\\src\\blog\\tests\\.hypothesis\\examples')
rootdir: G:\src\blog\tests, configfile: pytest.ini
plugins: allure-pytest-2.9.43, caterpillar-pytest-0.0.2, hypothesis-6.31.6,
forked-1.3.0, rerunfailures-10.1, xdist-2.3.0
collecting ... collected 9 items

<Package example>
  <Module test_example_01.py>
    <Function test_example>
  <Module test_example_02.py>
    <Function test_example>
  <Module test_example_03.py>
    <Function test_example>
<Package foobar>
  <Module test_foobar_01.py>
    <Function test_example>
  <Module test_foobar_02.py>
    <Function test_example>
  <Module test_foobar_03.py>
    <Function test_example>
<Package workd>
  <Module test_world_01.py>
    <Function test_example>
  <Module test_world_02.py>
    <Function test_example>
  <Module test_world_03.py>
    <Function test_example>

======================== 9 tests collected in 0.22s =======================
```

2.3.9 通过 Pytest.ini 设置用例默认的搜索目录

Pytest 命令默认是从当前目录开始搜索,但不建议修改这个规则的,因为 Pytest 是支持修改的,比如在一些特殊的情况下,当需要修改默认的搜索目录时,则可以通过 Pytest.ini 进行设置。

同样比如如下目录:

```
tests/
|-- pytest.ini
|-- example
|   |-- test_example_01.py
|   |-- test_example_02.py
|   '-- test_example_03.py
|-- foobar
|   |-- test_foobar_01.py
|   |-- test_foobar_02.py
|   '-- test_foobar_03.py
'-- hello
    '-- world
        |-- test_world_01.py
        |-- test_world_02.py
        '-- test_world_03.py
```

在某些特殊的场景下,如果 Pytst 只在 foobar 和 hello 目录下搜索脚本,而不能去其他目录下搜索,此时,就可以通过 Pytest.ini 中配置,如下所示。

```
[pytest]
testpaths = foobar hello
'''
```

执行结果如下,即指定了当执行 Pytest 命令的时候只会去搜索 foobar 和 hello 目录。

```bash
$ pytest
=========================== test session starts ===========================
platform win32 -- Python 3.9.6, pytest-6.2.5, py-1.10.0, pluggy-1.0.0
rootdir: G:\src\blog\tests, configfile: pytest.ini, testpaths: foobar, hello
plugins: allure-pytest-2.9.43, caterpillar-pytest-0.0.2, forked-1.3.0, rerunfailures-10.1, xdist-2.3.0
collected 6 items

foobar\test_foobar_01.py .                                          [ 16%]
foobar\test_foobar_02.py .                                          [ 33%]
```

```
foobar\test_foobar_03.py .                                          [ 50%]
hello\workd\test_world_01.py .                                      [ 66%]
hello\workd\test_world_02.py .                                      [ 83%]
hello\workd\test_world_03.py .                                      [100%]

============================ 6 passed in 0.05s ==========================
```

在Pytest.ini文件中,可以通过norecursedirs指定目录而不搜索Pytest.ini内容,如下所示。

```
[pytest]
norecursedirs = foobar hello
```

如下所示,foobar和hello目录明显都没有搜索。

```
$ pytest
============================ test session starts ==========================
platform win32 -- Python 3.9.6, pytest-6.2.5, py-1.10.0, pluggy-1.0.0
rootdir: G:\src\blog\tests, configfile: pytest.ini
plugins: allure-pytest-2.9.43, caterpillar-pytest-0.0.2, forked-1.3.0, rerun-faiures-10.1, xdist-2.3.0
collected 3 items

example\test_example_01.py .                                        [ 33%]
example\test_example_02.py .                                        [ 66%]
example\test_example_03.py .                                        [100%]

============================ 3 passed in 0.04s ==========================
```

2.3.10 在IDE中通过右键执行当前文件用例

若使用Pycharm编辑器编写自动化脚本,有时候希望像执行代码一样执行自动化脚本,就在文件中单击右键即可执行,但需要在测试脚本里加几行代码,在编辑器中单击右键执行。

```
import pytest


def test_func1():
    print("\nin test_case01.test_func1......")
    assert1 == 1


def test_func2():
    print("\nin test_case01.test_func2......")
```

```
    assert 1 == 1


if __name__ == "__main__":
    pytest.main()
```

指定 Pytest 的参数的情况如下所示。

```
import pytest


def test_func1():
    print("\nin test_case01.test_func1......")
    assert 1 == 1


def test_func2():
    print("\nin test_case01.test_func2......")
    assert 1 == 1


if __name__ == "__main__":
    pytest.main(['-s','test_case01.py'])
```

2.4　Pytest 脚本的加载原理

Pytest 测试脚本的加载原理实质上是模块的导入原理，Pytest 把每个测试脚本都作为一个 module 进行导入，导入的模式支持 prepend、append 和 importlib 三种模式，默认情况下是 prepend 模式。

2.4.1　prepend 模式

Pytest 默认的就是 prepend 模式，下面以如下的目录结构情况详细解析 prepared 模式下 Pytest 脚本的加载原理。

```
demo01/
    |----demo02/
        |----demo04/
            |----__init__.py
            |----test_demo01.py
        |----demo03/
            |----__init__.py
```

```
|----test_demo02.py
```

加载原理分析流程：

（1）Pytest 识别 test_demo01.py 文件后，从当前位置开始向上递归，找到 __init__.py 文件的目录。如果找不到，比如这里就是 demo04，但因为 demo02 中没有 __init__.py，所以从 test_demo01.py 开始找到的最上层带有 __init__.py 文件的目录是 demo04。

（2）此时 Pytest 把 demo04 的上一层目录，即 demo02 的目录路径插到 sys.path 的开头，则 prepend 就是表示从头插入。

（3）然后，开始计算导入模块的相对路径，比如这里是 demo04.test_demo01。

（4）将模块加入到 sys.modeules 中，sys.modules 是一个字典，key 为相对路径，比如这里是 demo04.test_demo01，value 是其对应的模块对象。

（5）Pytest 继续识别 test_demo02.py 文件，同样此时找到的 demo03 就是最顶层的带 __init__.py 的目录，之后把 demo03 的上一层目录，即 demo01 的目录插入到 sys.path 的头。

（6）同理，此时导入模块后将 demo03.test_demo02 加入到 sys.modules 中。

至此，Pytest 就把测试用例加载完成了。

test_demo01.py 和 test_demo02.py 内容均如下所示，为了演示加载原理，增加了打印 sys.path 和 sys.modules 的内容。

```python
import sys

print(f"sys.path:{sys.path}")
for elem in sys.modules.keys():
    if "demo" in elem:
        print(f"module:{elem}")

def test_func():
    assert1 == 1
```

从下面的执行结果可以看出，Pytest 首先把 'G:\\src\\blog\\tests\\demo01\\demo02' 插到 sys.path 的第一个元素，然后把 demo04.test_demo01 模块写到 sys.modeules 中，紧接着又把 'G:\\src\\blog\\tests\\demo01' 插到 sys.path 的第一个元素，再把 demo03.test_demo02 插到 sys.modules 中，与我们前面分析过程完全一致。

```
$ pytest -s
=============================== test session starts ========================
platform win32 -- Python 3.9.6, pytest-6.2.5, py-1.10.0, pluggy-1.0.0
rootdir: G:\src\blog\tests
plugins: allure-pytest-2.9.43, caterpillar-pytest-0.0.2, hypothesis-6.31.6, forked-1.3.0, rerunfailures-10.1, xdist-2.3.0
collecting ... sys.path:['G:\\src\\blog\\tests\\demo01\\demo02', 'D:\\python39\\Scripts\\pytest.exe',
```

```
'D:\\python39\\python39.zip', 'D:\\python39\\DLLs', 'D:\\python39\\
lib', 'D:\\python39', 'D:\\python39\\lib\\site-packages']
module:demo04
module:demo04.test_demo01
sys.path:['G:\\src\\blog\\tests\\demo01', 'G:\\src\\blog\\tests\\demo01\\demo02',
'D:\\python39\\Scripts\\pytest.exe', 'D:\\python39\\python39.zip', 'D:\\python39\\DLLs'
, 'D:\\python39\\lib', 'D:\\python39', 'D:\\python39\\lib\\site-packages']
module:demo04
module:demo04.test_demo01
module:demo03
module:demo03.test_demo02
collected 2 items

demo01\demo02\demo04\test_demo01.py .
demo01\demo03\test_demo02.py .

============================== 2 passed in 0.08s ==========================
```

2.4.2 append 模式

append 模式的整个流程与 prepend 模式是完全一样的，唯一的区别就是在将找到的目录插入 sys.path 的时候，append 是插到 sys.path 的末尾，prepend 则是插到 sys.path 的开头。大家可以通过 import-mode=append 来指定导入模式为 append，执行结果如下所示。可以看出，这时路径已经插到 sys.path 的末尾了，这一点与 prepend 是不同的。

```
$ pytest -s --import-mode=append
============================== test session starts ========================
platform win32 -- Python 3.9.6, pytest-6.2.5, py-1.10.0, pluggy-1.0.0
rootdir: G:\src\blog\tests
plugins: allure-pytest-2.9.43, caterpillar-pytest-0.0.2, hypothesis-6.31.6,
forked-1.3.0, rerunfailures-10.1, xdist-2.3.0
collecting ... sys.path:['D:\\python39\\Scripts\\pytest.exe', 'D:\\python39\\python39.
zip', 'D:\\python39\\DLLs', 'D:\\python39\\lib', 'D:\\python39', 'D:\\python39\\lib
\\site-packages', 'G:\\src\\blog\\tests\\demo01\\demo02']
module:demo04
module:demo04.test_demo01
sys.path:['D:\\python39\\Scripts\\pytest.exe', 'D:\\python39\\python39.zip', 'D:\\py-
thon39\\DLLs', 'D:\\python39\\lib', 'D:\\python39', 'D:\\python39\\lib\\site-packages
', 'G:\\src\\blog\\tests\\demo01\\demo02', 'G:\\src\\blog\\tests\\demo01']
module:demo04
module:demo04.test_demo01
module:demo03
```

```
module:demo03.test_demo02
collected 2 items

demo01\demo02\demo04\test_demo01.py .
demo01\demo03\test_demo02.py .

=========================== 2 passed in 0.04s ===========================
```

2.4.3　prepend 和 append 模式存在的问题

prepend 和 append 模式都存在一个问题,那就是需要保持导入模块的唯一性,如要解释这个问题前我们先看一个例子,目录结构情况如下所示。

```
demo01/
    |----demo02/
        |----demo04/
            |----__init__.py
            |----test_demo01.py
    |----demo04/
        |----__init__.py
        |----test_demo01.py
```

我们根据上面的导入原理分析一下,不论是 prepend 模式还是 append 模式,最终两个 test_demo01.py 要导入的模块名都是 demo01.test_demo01。在导入这两个模块后,将它们写入 sys.modules 时会报错,因为 sys.modules 是字典类型的,字典类型的 key 是不允许重复的。

两个 test_demo01.py 的代码如下所示。

```
import sys

print(f"sys.path:{sys.path}")
for elem in sys.modules.keys():
    if "demo" in elem:
        print(f"module:{elem}")


def test_func():
    assert1 == 1
```

执行结果与前面分析结果是一致的,换言之,如果执行 Pytest 的时候出现了如下错误,那么说明这个导入模块重名了。

```
$ pytest -s
=========================== test session starts ===========================
platform win32 -- Python 3.9.6, pytest-6.2.5, py-1.10.0, pluggy-1.0.0
```

```
rootdir: G:\src\blog\tests
plugins: allure-pytest-2.9.43, caterpillar-pytest-0.0.2, hypothesis-6.31.6,
forked-1.3.0, rerunfailures-10.1, xdist-2.3.0
collecting ... sys.path:['G:\\src\\blog\\tests\\demo01\\demo02', 'D:\\python39\\Scripts\\pytest.exe',
'D:\\python39\\python39.zip', 'D:\\python39\\DLLs', 'D:\\python39\\lib', 'D:\\python39',
'D:\\python39\\lib\\site-packages']
module:demo04
module:demo04.test_demo01
collected 1 item / 1 error

================================= ERRORS =================================
_____ ERROR collecting demo01/demo04/test_demo01.py _____
import file mismatch:
imported module 'demo04.test_demo01' has this __file__ attribute:
  G:\src\blog\tests\demo01\demo02\demo04\test_demo01.py
which is not the same as the test file we want to collect:
  G:\src\blog\tests\demo01\demo04\test_demo01.py
HINT: remove __pycache__ / .pyc files and/or use a unique basename for your test file modules
========================= short test summary info =========================
ERROR demo01/demo04/test_demo01.py
!!!!!!!!!!!!!!!!!!!!!!!!! Interrupted: 1 error during collection !!!!!!!!!!!!!!!!!!!!!!!!!
============================= 1 error in 0.16s =============================
```

解决这个问题比较简单的方法就是，在每个文件夹中都加一个 __init__.py 文件，如下所示。

```
demo01/
    |----__init__.py
    |----demo02/
        |__init__.py
        |----demo04/
            |----__init__.py
            |----test_demo01.py
    |----demo04/
        |----__init__.py
        |----test_demo01.py
```

这样来分析一下：从第一个 test_demo01.py 往上找，发现 demo01 是最后一个带 __init__.py 的文件夹，我们把 demo01 的上一层目录加到 sys.path，此时，第一个 test_demo01.py 的导入模块就变为 demo01.demo02.demo04.test_demo01。同理第二个 test_demo01.py 的导入模块就变为 demo01.demo04.test_demo01，这样就解决了执行

错误的问题。

也正是这个原因,许多文章或者教程中讲 Pytest 要求文件夹必须带 __init__.py,甚至有的宣称如果不加 __init__.py 不会被识别,这是不准确的。为了减少麻烦,可以让新建文件夹都直接带上 __init__.py 文件,以保证不会出执行错误问题的。

2.4.4　importlib 模式

importlib 模式是 Pytest6.0 以后版本支持的新方式,importlib 方式不再需要修改 sys.path 和 sys.modules,因此不存在上面 prepend 和 append 面临的潜在的问题,这里我们采用一种全新的导入方式,首先看一个例子目录结构,如下所示。

```
demo01/
    |----demo02/
         |----demo04/
              |----test_demo01.py
    |----demo04/
         |----test_demo01.py
```

如果按照 prepend 或 append 的思路分析,是执行不起来的,导入模块的名字是重复的,我们可以执行如下代码。

```
$ pytest -s
=========================== test session starts ===========================
platform win32 -- Python 3.9.6, pytest-6.2.5, py-1.10.0, pluggy-1.0.0
rootdir: G:\src\blog\tests
plugins: allure-pytest-2.9.43, caterpillar-pytest-0.0.2, hypothesis-6.31.6,
forked-1.3.0, rerunfailures-10.1, xdist-2.3.0
collecting ... sys.path:['G:\\src\\blog\\tests\\demo01\\demo02\\demo04', 'D:\\python39
\\Scripts\\pytest.exe', 'D:\\python39\\python39.zip', 'D:\\python39\\DLLs', 'D:\\py
thon39\\lib', 'D:\\python39', 'D:\\python39\\lib\\site-packages']
module:test_demo01
collected 1 item / 1 error


================================= ERRORS =================================
_____ ERROR collecting demo01/demo04/test_demo01.py _____
import file mismatch:
imported module 'test_demo01' has this __file__ attribute:
    G:\src\blog\tests\demo01\demo02\demo04\test_demo01.py
which is not the same as the test file we want to collect:
    G:\src\blog\tests\demo01\demo04\test_demo01.py
HINT: remove __pycache__ / .pyc files and/or use a unique basename for your test file modules
========================== short test summary info ==========================
ERROR demo01/demo04/test_demo01.py
!!!!!!!!!!!!!!!!!!!!!!!!!!! Interrupted: 1 error during collection !!!!!!!!!!!!!!!!!!!!!!
```

```
============================= 1 error in 0.16s =============================
```

importlib 模式不会去修改 sys.paht 和 sys.mo,因此,也就不会有执行错误问题了,执行结果如下所示。

```
$ pytest -s --import-mode=importlib
============================= test session starts =============================
platform win32 -- Python 3.9.6, pytest-6.2.5, py-1.10.0, pluggy-1.0.0
rootdir: G:\src\blog\tests
plugins: allure-pytest-2.9.43, caterpillar-pytest-0.0.2, hypothesis-6.31.6, forked-1.3.0, rerunfailures-10.1, xdist-2.3.0
collecting ... sys.path:['D:\\python39\\Scripts\\pytest.exe', 'D:\\python39\\python39.zip', 'D:\\python39\\DLLs', 'D:\\python39\\lib', 'D:\\python39', 'D:\\python39\\lib\\site-packages']
sys.path:['D:\\python39\\Scripts\\pytest.exe', 'D:\\python39\\python39.zip', 'D:\\python39\\DLLs', 'D:\\python39\\lib', 'D:\\python39', 'D:\\python39\\lib\\site-packages']
collected 2 items

demo01\demo02\demo04\test_demo01.py .
demo01\demo04\test_demo01.py .

============================= 2 passed in 0.06s =============================
```

这就是 Pytest 自动化脚本的加载原理,至此,也就明白当前 Pytest 默认情况下采用的是 prepared 模式,而在这种模式下,如果文件夹中没有 __init__.py 文件,一定要保持测试文件命名的独一无二性,所以在实践中,为了减少一些潜在的问题,建议在创建文件夹的时候,直接在所有文件夹下创建 __init__.py 文件,如此则不需要担心测试脚本文件名重名的问题了。

第 3 章 Assert 断言

在测试领域,断言就是对执行脚本的结果与预期结果进行比较,当断言为 True 时表示用例通过,当断言为 False 时,表示用例失败。Pytest 框架中断言直接使用 Assert 语句,Pytest 本身就是基于 Python 语言的,Pytest 中断言就是判断 Assert 语句后面的表达式是否为 True,如果 Assert 语句后面的表达式为 True 表示断言通过,用例 pass,如果断言后面的表达式为 False,表示断言失败,即测试用例失败。因此,断言的核心对 Python 中 True 和 False 的理解。

3.1 使用 Assert 断言

3.1.1 Python 中为 False 的数值断言均失败

在 Python 语言中,表示 False 的数值即各个数据类型的空值,如整数的 0,字符串的空字符串,None 值、空列表、空字典、空集合、空元组,布尔类型的 False 等,这里需要注意,一个空格的字符串的值不是 False,编写测试脚本如下所示。

```python
def test_demo01():
    assert 0
def test_demo02():
    assert 0.0
def test_demo03():
    assert ""
def test_demo04():
    assert ()
def test_demo05():
    assert {}
def test_demo06():
    assert []
def test_demo07():
    assert None
def test_demo08():
    assert False
def test_demo09():
    assert " "
```

从执行结果中可以看出有 8 个用例失败,有 1 个用例通过,通过的那个用例即会对空格字符串作出判断,即空格字符串的布尔值是 True,而空字符串的布尔值则是 False。

```
$ pytest
============================ test session starts =============================
platform win32 -- Python 3.9.13, pytest-7.1.2, pluggy-1.0.0
rootdir: G:\redrose2100_book\ebooks\Pytest 企业级应用实战\src
collected 9 items

test_demo.py FFFFFFFF.                                                 [100%]

================================== FAILURES ==================================
_____ test_demo01 _____

    def test_demo01():
>       assert 0
E       assert 0

test_demo.py:2: AssertionError
_____ test_demo02 _____

    def test_demo02():
>       assert 0.0
E       assert 0.0

test_demo.py:5: AssertionError
_____ test_demo03 _____

    def test_demo03():
>       assert ""
E       AssertionError: assert ''

test_demo.py:8: AssertionError
_____ test_demo04 _____

    def test_demo04():
>       assert ()
E       assert ()

test_demo.py:11: AssertionError
_____ test_demo05 _____
```

```
    def test_demo05():
>       assert {}
E       assert {}

test_demo.py:14: AssertionError
_____ test_demo06 _____

    def test_demo06():
>       assert []
E       assert []

test_demo.py:17: AssertionError
_____ test_demo07 _____

_____ test_demo08 _____

    def test_demo08():
>       assert False
E       assert False

test_demo.py:23: AssertionError
=========================== short test summary info ====================
FAILED test_demo.py::test_demo01 - assert 0
FAILED test_demo.py::test_demo02 - assert 0.0
FAILED test_demo.py::test_demo03 - AssertionError: assert ''
FAILED test_demo.py::test_demo04 - assert ()
FAILED test_demo.py::test_demo05 - assert {}
FAILED test_demo.py::test_demo06 - assert []
FAILED test_demo.py::test_demo07 - assert None
FAILED test_demo.py::test_demo08 - assert False
========================= 8 failed, 1 passed in 0.16s ==================
```

3.1.2 Python 逻辑表达式为 False 的断言均失败

在 Python 中,常用的逻辑判断表达式,比如==、!=、<、>、<=、>=、is、is not、in、not in、is None、is not None 等,均会断言时失败,编写测试用例代码如下所示。

```
def test_demo01():
    assert 1 == 2


def test_demo02():
    assert 3 < 2
```

```
def test_demo03():
    assert 3 is str

def test_demo04():
    assert 3 in [1,2,4]

def test_demo05():
    assert 0 is None
```

即五个测试用例均失败执行结果如下所示。

```
$ pytest
=========================== test session starts ===========================
platform win32 -- Python 3.9.13, pytest-7.1.2, pluggy-1.0.0
rootdir: G:\redrose2100_book\ebooks\Pytest 企业级应用实战\src
collected 5 items

test_demo.py FFFFF                                                  [100%]

================================= FAILURES ================================
_____ test_demo01 _____

    def test_demo01():
>       assert 1 == 2
E       assert 1 == 2

test_demo.py:2: AssertionError
_____ test_demo02 _____

    def test_demo02():
>       assert 3 < 2
E       assert 3 < 2

test_demo.py:5: AssertionError
_____ test_demo03 _____

    def test_demo03():
test_demo.py:11: AssertionError
_____ test_demo05 _____

    def test_demo05():
>       assert 0 is None
E       assert 0 is None
```

```
test_demo.py:14: AssertionError
========================= short test summary info =========================
FAILED test_demo.py::test_demo01 - assert 1 == 2
FAILED test_demo.py::test_demo02 - assert 3 < 2
FAILED test_demo.py::test_demo03 - assert 3 is str
FAILED test_demo.py::test_demo04 - assert 3 in [1, 2, 4]
FAILED test_demo.py::test_demo05 - assert 0 is None
========================= 5 failed in 0.14s =========================
```

如上所述,在 Pytest 中,断言 Assert 语句的使用就是这么简单,主要看 Assert 后的数值或者表达式的布尔值。如果为 True,则用例通过;如果为 False,则用例失败。

3.2 自定义断言报错信息

在自定义断言报错信息之前,首先观察一下默认情况的报错信息。

```
def test_demo():
    a = 5
    assert a == 10
```

从执行结果中仔细观察也是能看出来错误原因的,只不过我们测试人员通常追求极简,希望直接就能看出错误的原因。

```
$ pytest
========================= test session starts =========================
platform win32 -- Python 3.9.13, pytest-7.1.2, pluggy-1.0.0
rootdir: G:\redrose2100_book\ebooks\Pytest 企业级应用实战\src
collected 1 item

test_demo.py F                                                  [100%]

=============================== FAILURES ==============================
_____ test_demo _____

    def test_demo():
        a = 5
>       assert a == 10
E       assert 5 == 10

test_demo.py:3: AssertionError
========================= short test summary info =========================
FAILED test_demo.py::test_demo - assert 5 == 10
========================= 1 failed in 0.11s =========================
```

Pytest的Assert断言语句是提供了自定义断言报错信息的功能的,使用起来也是非常的简单,只要在Assert断言语句后紧跟着加上一句错误信息,这样当断言成功时,增加的信息就不会打印,而如果断言失败时,则增加的断言信息就会打印出来,这样一来当测试人员看到自己增加的断言错误信息描述,就能做到一眼就能看出来错误的原因了。

编写的代码如下所示。

```python
def test_demo():
    a = 5
    assert a == 10 , "变量a期望是10,当前为%d" % a
```

从上述执行结果中,可以很快从断言报错信息中看出断言失败的原因,即"变量a期望的值是10,当前为5",如此以来,对自动化脚本调试和执行时定位问题等非常方便。

```
$ pytest
============================ test session starts ============================
platform win32 -- Python 3.9.13, pytest-7.1.2, pluggy-1.0.0
rootdir: G:\redrose2100_book\ebooks\Pytest企业级应用实战\src
collected 1 item

test_demo.py F                                                         [100%]

================================== FAILURES =================================
_____ test_demo _____

    def test_demo():
        a = 5
>       assert a == 10 , "变量a期望是10,当前为%d" % a
E       AssertionError: 变量a期望是10,当前为5
E       assert 5 == 10

test_demo.py:3: AssertionError
========================== short test summary info ==========================
FAILED test_demo.py::test_demo - AssertionError: 变量a期望是10,当前为5
============================= 1 failed in 0.11s =============================
```

3.3 对捕获的异常进行断言

3.3.1 对异常类型进行断言

对捕获的异常进行断言主要用于,比如异常测试时,当我们给定了特殊的数据时,程序如果产生了我们期望的异常,那么对测试而言是正确的,此时就可以通过对捕获的异常类型进行断言。当然,对捕获的异常类型进行断言完全可以使用 Python 语言中的 try…except 结构进行判断,比如对一个除法函数,当我们给定除数为 0 时,那么程序报除数为 0 的异常是正确的,使用 Python 语法中的 try…except 结构实现的测试代码如下所示。

```
def div(a,b):
    return a/b

def test_demo():
    try:
        c = div(10,0)
    except Exception as e:
        assert isinstance(e,ZeroDivisionError)
```

执行结果如下所示,即表示此用例通过。

```
$ pytest
=========================== test session starts ===========================
platform win32 -- Python 3.9.13, pytest-7.1.2, pluggy-1.0.0
rootdir: D:\redrose2100-book\ebooks\Pytest 企业级应用实战\src
collected 1 item

test_demo.py .                                                     [100%]

=========================== 1 passed in 0.07s ===========================
```

从上面的测试代码中也看得出,通过 try…except 结构对异常类型进行断言有些烦琐,Pytest 提供了一种更加便捷的方式,即使用 with pytest.raise(异常类型)的格式,如下所示。

```
import pytest

def div(a,b):
    return a/b

def test_demo():
    with pytest.raises(ZeroDivisionError):
        c = div(10,0)
```

从上述执行结果可以看出用例通过了,此时,异常类型确实与我们期望的是一致的。

```
$ pytest
=========================== test session starts ===========================
```

```
platform win32 -- Python 3.9.13, pytest-7.1.2, pluggy-1.0.0
rootdir: D:\redrose2100-book\ebooks\Pytest企业级应用实战\src
collected 1 item

test_demo.py .                                                   [100%]

=========================== 1 passed in 2.11s ===========================
```

3.3.2 对捕获的异常信息进行断言

对捕获的异常信息进行断言，同样可以使用 Python 语言中的 try…except 结构。

```
def div(a,b):
    return a/b

def test_demo():
    try:
        c = div(10,0)
    except Exception as e:
        assert "division by zero" in str(e)
```

从执行结果可以看得出，此时，捕获的异常的信息与期望是一致的。

```
$ pytest
=========================== test session starts ===========================
platform win32 -- Python 3.9.13, pytest-7.1.2, pluggy-1.0.0
rootdir: D:\redrose2100-book\ebooks\Pytest企业级应用实战\src
collected 1 item

test_demo.py .                                                   [100%]

=========================== 1 passed in 2.03s ===========================
```

虽然上面的代码可以对异常的信息进行断言，但明显过于烦琐，Pytest 框架也提供了一种比较方便的方式可用于对捕获到的异常信息进行断言。通过使用 with Pytest.raise(异常类型)ase，此时异常的信息均在对象 e 的 value 属性中，即断言期望的异常信息在 e.value 中。

```
import pytest

def div(a,b):
    return a/b
def test_demo():
    with pytest.raises(ZeroDivisionError) as e:
        c = div(10,0)
```

```
assert "division by zero" in str(e.value)
```

从执行结果可以看出，此时异常信息中包含 division by zero 字符串。

```
$ pytest
=========================== test session starts ===========================
platform win32 -- Python 3.9.13, pytest-7.1.2, pluggy-1.0.0
rootdir: D:\redrose2100-book\ebooks\Pytest 企业级应用实战\src
collected 1 item

test_demo.py .                                                      [100%]

============================ 1 passed in 2.02s ============================
```

3.3.3 同时对捕获的异常类型和异常信息进行断言

我们同样不厌其烦地使用 Python 的语法中的 try...except 结构对捕获到的异常类型和异常信息进行断言，大家别小看了这些基本的或看上去不怎么高明的手段，其实很多时候在工作中，当我们并未掌握一个框架或者工具提供的巧妙的方法，并不能说明我们不能工作了，而完全可以采用我们已掌握的方法的达到同样的目的。我们这里采用了两次断言分别对异常类型和异常信息进行了断言。

```
def div(a,b):
    return a/b

def test_demo():
    try:
        c = div(10,0)
    except Exception as e:
        assert isinstance(e, ZeroDivisionError)
        assert "division by zero" in str(e)
```

从下面执行结果可以看出，此时是完全可以对异常类型和异常信息进行断言的。

```
$ pytest
=========================== test session starts ===========================
platform win32 -- Python 3.9.13, pytest-7.1.2, pluggy-1.0.0
rootdir: D:\redrose2100-book\ebooks\Pytest 企业级应用实战\src
collected 1 item

test_demo.py .                                                      [100%]

============================ 1 passed in 2.10s ============================
```

同样，Pytest 也提供了一种更加简便的方法，即通过指定异常类型以及通过正则表达式来匹配异常信息的方式同时对异常类型和异常信息进行断言。

```python
import pytest


def div(a,b):
    return a/b


def test_demo():
    with pytest.raises(ZeroDivisionError,match=r"division\s*by\s*zero"):
        c = div(10,0)
```

执行结果显示用例通过，捕获的异常类型是ZeroDivisionError，而且异常信息也能匹配正则表达式"division\s*by\s*zero"，这样整个代码就显得整洁多了。而且用起来也非常方便。

```
$ pytest
============================ test session starts ============================
platform win32 -- Python 3.9.13, pytest-7.1.2, pluggy-1.0.0
rootdir: D:\redrose2100-book\ebooks\Pytest企业级应用实战\src
collected 1 item

test_demo.py .                                                        [100%]

============================ 1 passed in 2.00s =============================
```

3.3.4 对一个函数可能产生的异常进行断言

通过Pytest.raise还可以对一个函数可能产生的异常进行断言，在格式的调用中，会执行func函数，并将args和**kwargs传入func函数，然后判断此时func中报出的异常是否与第一个参数异常一致。

```
pytest.raises(ExpectedException, func, args, **kwargs)
```

如果两个用例分别去调用除法计算函数div，则均断言会产生除数为零的异常。

```python
import pytest


def div(a,b):
    return a/b


def test_demo01():
    pytest.raises(ZeroDivisionError,div,10,0)
```

```
def test_demo02():
    pytest.raises(ZeroDivisionError,div,10,5)
```

执行结果显示如下，显然 test_demo02 中不会产生除数为零的异常，因此此用例会失败。

```
$ pytest
=========================== test session starts ===========================
platform win32 -- Python 3.9.13, pytest-7.1.2, pluggy-1.0.0
rootdir: D:\redrose2100-book\ebooks\Pytest 企业级应用实战\src
collected 2 items

test_demo.py .F                                                    [100%]

================================= FAILURES ================================
_____ test_demo02 _____

    def test_demo02():
>       pytest.raises(ZeroDivisionError,div,10,5)
E       Failed: DID NOT RAISE <class 'ZeroDivisionError'>

test_demo.py:10: Failed
========================= short test summary info =========================
FAILED test_demo.py::test_demo02 - Failed: DID NOT RAISE <class 'ZeroDivisionError'>
======================== 1 failed, 1 passed in 2.14s ======================
```

3.4 重写断言 Assert 语句的报错信息

3.4.1 默认的报错信息

在如下测试代码中，这里面有两个测试用例，一个是判断 1 是否等于 2，第二个是判断字符串 "hello" 是否等于整数 10。

```
def test_demo01():
    assert 1 == 2

def test_demo02():
    assert "hello" == 10
```

执行结果如下所示。

```
$ pytest
```

```
=========================== test session starts ===========================
platform win32 -- Python 3.9.13, pytest-7.1.2, pluggy-1.0.0
rootdir: D:\redrose2100-book\ebooks\Pytest企业级应用实战\src
collected 2 items

test_demo.py FF                                                      [100%]

================================= FAILURES ================================
_____ test_demo02 _____

    def test_demo02():
>       assert "hello" == 10
E       AssertionError: assert 'hello' == 10

test_demo.py:5: AssertionError
========================= short test summary info =========================
FAILED test_demo.py::test_demo01 - assert 1 == 2
FAILED test_demo.py::test_demo02 - AssertionError: assert 'hello' == 10
============================ 2 failed in 2.24s ============================
```

对于许多测试人员，可能习惯于看比较准确的中文描述，当大家看到报错时，不管报错描述多么很清晰，从多年的自动化测试经历中发现，很多测试人员或者编写自动化测试脚本的测试人员，在调试或者执行自动化用例的时候仍然很茫然，他们更加希望看到更加人性化的中文描述，当然 Pytest 的断言也提供了在脚本中自定义报错信息的方式，如下所示。

```
def test_demo01():
    assert 1 == 2,"期望值与实际值不一致,期望值为2,实际值为1"
```

但是我们在实际测试开发中发现，虽然上述方式给自动化脚本开发人员自己定义的权利，但如果在进行脚本开发的过程中，每个断言都增加断言报错信息，又很占用时间，就会不方便。有很多断言报错信息其实都是类似的，比如判断两个变量是否相等。其实断言信息都是类似的，而在脚本中，每次断言都加断言报错信息又会显得很冗余，因此，另外一种在 conftest.py 中重写断言报错信息的处理方式就显得非常好用了。

3.4.2　重写判断是否相等的断言报错信息

我们先使用一个 conftest.py 文件，这个文件的详细的作用以及使用情况在后面我们会继续展开讲解，这里仅仅需要知道当 conftest.py 文件和测试用例文件放在同一个目录时，在执行测试用例文件之前会自动先加载执行 conftest.py，因此本小节正是利用这一点，在 conftest.py 文件中重写 pytest_assertrepr_compare 方法。

我们看 conftest.py 文件中的代码,在进行判断两个变量是否相等时,首先判断一下这两个变量是否为同一类型,如果不是同一类型,直接报错,并在报错中提示类型错误。这种情况比较常见的是:在判断两个变量是否相等时,期望值为数字1,而实际值为字符串1,当打印调试的时候,又会发现没有区别,但是结果却是报错。这种情况很是困扰人,因此可以在 pytest_assertrepr_compare 直接重写,并判断,当类型一致时,等再进行是否相等的判断,如果不相等,则采用中文详细描述。

```python
def pytest_assertrepr_compare(op, left, right):
    if op == "==":
        if not isinstance(right,type(left)):
            return [
                f"断言错误,期望值与实际值的数据类型不一致,期望值数据类型为:{type(right)},实际值为:{type(left)}",
                f"期望值为:{right},实际值为:{left}",
            ]
        else:
            return [
                f"断言错误,期望值与实际值不一致,期望值为:{right},实际值为:{left}",
            ]
```

之后在同一目录下创建 test_demo.py 文件,测试脚本如下所示。

```python
def test_demo01():
    assert 1 == 2

def test_demo02():
    assert "10" == 10
```

从执行结果中可以看到,此时错误描述非常简单、清晰。

```
$ pytest
========================= test session starts =========================
platform win32 -- Python 3.9.13, pytest-7.1.2, pluggy-1.0.0
rootdir: D:\redrose2100-book\ebooks\Pytest企业级应用实战\src
collected 2 items

test_demo.py FF                                                  [100%]


=============================== FAILURES ===============================
_____ test_demo01 _____

    def test_demo01():
>       assert 1 == 2
E       assert 断言错误,期望值与实际值不一致,期望值为:2,实际值为:1
```

```
test_demo.py:2: AssertionError
_____ test_demo02 _____

    def test_demo02():
>       assert "10" == 10
E       AssertionError: assert 断言错误,期望值与实际值的数据类型不一致,期望值数据类型为:<class 'int'>，实际值为:<class 'str'>
E         期望值为:10,实际值为:10

test_demo.py:5: AssertionError
========================== short test summary info ==========================
FAILED test_demo.py::test_demo01 - assert 断言错误,期望值与实际值不一致,期望值为:2,实际值为:1
FAILED test_demo.py::test_demo02 - AssertionError: assert 断言错误,期望值与实际值的数据类型不一致,期望值数据类型为:<class 'int'>，实际值为:<class 'str'>
============================ 2 failed in 0.15s =============================
```

像上述这种细节处理,就能很好地显示出自动化测试的专业性了,资深的自动化测试专家设计出的自动化测试框架优点,很多时候都是体现在这些细节上。

3.4.3　重写常见的判断逻辑报错信息

我们重写==,in,not in 的断言报错信息,其他的比如 >,<,>=,<=,!=等运算符重写方法类似,在企业级实战中,设计自动化测试框架的时候,要尽量重写全面一些,这里不再一一列举,有兴趣的读者可以自己尝试将这些运算符补全,练习在conftest.py 中编写代码。

```python
def pytest_assertrepr_compare(op, left, right):
    if op == "==":
        if not isinstance(right,type(left)):
            return [
                f"断言错误,期望值与实际值的数据类型不一致,期望值数据类型为:{type(right)},实际值为:{type(left)}",
                f"期望值为:{right},实际值为:{left}",
            ]
        else:
            return [
                f"断言错误,期望值与实际值不一致,期望值为:{right},实际值为:{left}",
            ]
    if op == "in":
        if isinstance(left,str) and isinstance(right,str):
```

```
            return [
                    f"期望 {left} 是 {right} 的子串,实际 {left} 不是 {right} 的子串,"
            ]
        elif isinstance(right,list) or isinstance(right,set) or isinstance(right,tuple):
            return [
                    f"期望 {left} 是集合 {right} 中的一个元素,实际集合 {right} 中没有 {left} 元素"
            ]
        elif isinstance(right,dict):
            return [
                    f"期望 {left} 是字典 {right} 中的一个 key,实际字典 {right} 中没有值为 {left} 的 key"
            ]
        else:
            return [
                    f"期望 {left} 是 {right} 中的一部分,实际上 {left} 并不是 {right} 的一部分"
            ]
    if op == "not in":
        if isinstance(left, str) and isinstance(right, str):
            return [
                    f"期望 {left} 不是 {right} 的子串,实际 {left} 是 {right} 的子串,"
            ]
        elif isinstance(right, list) or isinstance(right, set) or isinstance(right, tuple):
            return [
                    f"期望 {left} 不是集合 {right} 中的一个元素,实际集合 {right} 中有 {left} 元素"
            ]
        elif isinstance(right, dict):
            return [
                    f"期望 {left} 不是字典 {right} 中的一个 key,实际字典{right}中有值为 {left} 的 key"
            ]
        else:
            return [
                    f"期望 {left} 不是 {right} 中的一部分,实际上 {left} 是 {right} 的一部分"
            ]
```

我们在下面同级目录下编写了几个测试用例,用于演示上述重写的效果,具体代码如下所示。

```
def test_01():
    assert1 == 1
def test_02():
```

```
        assert 1 == 2
    def test_03():
        assert "1" == 1
    def test_04():
        assert "aa" in "bbaa"
    def test_05():
        assert "aa" in "bba"
    def test_06():
        assert "aa" in ["aa","bb"]
    def test_07():
        assert "aa" in ("aa","bb")
    def test_08():
        assert "aa" in {"aa","bb"}
    def test_09():
        assert "ab" in ["aa", "bb"]
    def test_10():
        assert "ab" in ("aa", "bb")
    def test_11():
        assert "ab" in {"aa", "bb"}
    def test_12():
        assert "name" in {"name":"张三丰","age":100}
    def test_13():
        assert "gender" in {"name":"张三丰","age":100}
    def test_14():
        assert "aa" not in "bbaa"
    def test_15():
        assert "aa" not in "bba"
    def test_16():
        assert "aa" not in ["aa","bb"]
    def test_17():
        assert "aa" not in ("aa","bb")
    def test_18():
        assert "aa" not in {"aa","bb"}
    def test_19():
        assert "ab" not in ["aa", "bb"]
    def test_20():
        assert "ab" not in ("aa", "bb")
    def test_21():
        assert "ab" not in {"aa", "bb"}
    def test_22():
        assert "name" not in {"name":"张三丰","age":100}
    def test_23():
        assert "gender" not in {"name":"张三丰","age":100}
```

从执行结果中可以看出，经过这么重写之后，断言报错信息基本一眼就看出是什么问题，当然这里只是举了个例子，在设计自动化测试框架的时候，应完全根据具体业务进行自定义,力求做到让团队或项目成员能最快、最简单、最容易地定位出错误。

```
$ pytest
============================ test session starts ============================
platform win32 -- Python 3.9.13, pytest-7.1.2, pluggy-1.0.0
rootdir: D:\redrose2100-book\ebooks\Pytest 企业级应用实战\src
collected 23 items

test_demo.py .FF.F...FFF.FF.FFF...F.                                  [100%]

================================== FAILURES =================================
_____ test_02 _____

    def test_02():
>       assert 1 == 2
E       assert 断言错误,期望值与实际值不一致,期望值为:2,实际值为:1

test_demo.py:4: AssertionError
_____ test_03 _____

    def test_03():
>       assert "1" == 1
E       AssertionError: assert 断言错误,期望值与实际值的数据类型不一致,期望值数据
类型为:<class 'int'>,实际值为:<class 'str'>
E         期望值为:1,实际值为:1

test_demo.py:6: AssertionError
_____ test_05 _____

    def test_05():
>       assert "aa" in "bba"
E       assert 期望 aa 是 bba 的子串,实际 aa 不是 bba 的子串,

test_demo.py:10: AssertionError
_____ test_09 _____

    def test_09():
>       assert "ab" in ["aa", "bb"]
E       AssertionError: assert 期望 ab 是集合 ['aa', 'bb'] 中的一个元素,实际集合
['aa', 'bb'] 中没有 ab 元素

test_demo.py:18: AssertionError
_____ test_10 _____
```

77

```
    def test_10():
>       assert "ab" in ("aa", "bb")
E       AssertionError: assert 期望 ab 是集合 ('aa', 'bb') 中的一个元素,实际集合 ('aa', 'bb') 中没有 ab 元素
```

test_demo.py:20: AssertionError
_____ test_11 _____

```
    def test_11():
>       assert "ab" in {"aa", "bb"}
E       AssertionError: assert 期望 ab 是集合 {'bb', 'aa'} 中的一个元素,实际集合 {'bb', 'aa'} 中没有 ab 元素
```

test_demo.py:22: AssertionError
_____ test_13 _____

```
    def test_13():
>       assert "gender" in {"name":"张三丰","age":100}
E       AssertionError: assert 期望 gender 是字典 {'name': '张三丰', 'age': 100} 中的一个 key,实际字典 {'name': '张三丰', 'age': 100} 中没有值为 gender 的 key
```

test_demo.py:26: AssertionError
_____ test_14 _____

```
    def test_14():
>       assert "aa" not in "bbaa"
E       assert 期望 aa 不是 bbaa 的子串,实际 aa 是 bbaa 的子串,
```

test_demo.py:28: AssertionError
_____ test_16 _____

```
    def test_16():
>       assert "aa" not in ["aa","bb"]
E       AssertionError: assert 期望 aa 不是集合 ['aa', 'bb'] 中的一个元素,实际集合 ['aa', 'bb'] 中有 aa 元素
```

test_demo.py:32: AssertionError
_____ test_17 _____

```
    def test_17():
>       assert "aa" not in ("aa","bb")
E       AssertionError: assert 期望 aa 不是集合 ('aa', 'bb') 中的一个元素,实际集合
```

('aa', 'bb') 中有 aa 元素

```
        test_demo.py:34: AssertionError
_____ test_18 _____

        def test_18():
>           assert "aa" not in {"aa","bb"}
E           AssertionError: assert 期望 aa 不是集合 {'bb', 'aa'} 中的一个元素,实际集合 {'bb', 'aa'} 中有 aa 元素

        test_demo.py:36: AssertionError
_____ test_22 _____

        def test_22():
>           assert "name" not in {"name":"张三丰","age":100}
E           AssertionError: assert 期望 name 不是字典 {'name': '张三丰', 'age': 100} 中的一个 key,实际字典{'name': '张三丰', 'age': 100}中有值为 name 的 key

        test_demo.py:44: AssertionError
========================= short test summary info =========================
FAILED test_demo.py::test_02 - assert 断言错误,期望值与实际值不一致,期望值为:2,实际值为:1
FAILED test_demo.py::test_03 - AssertionError: assert 断言错误,期望值与实际值的数据类型不一致,期望值数据类型为:<class 'int'>,实际值为:<class 'str'>
FAILED test_demo.py::test_05 - assert 期望 aa 是 bba 的子串,实际 aa 不是 bba 的子串,
FAILED test_demo.py::test_09 - AssertionError: assert 期望 ab 是集合 ['aa', 'bb'] 中的一个元素,实际集合 ['aa', 'bb'] 中没有 ab 元素
FAILED test_demo.py::test_10 - AssertionError: assert 期望 ab 是集合 ('aa', 'bb') 中的一个元素,实际集合 ('aa', 'bb') 中没有 ab 元素
FAILED test_demo.py::test_11 - AssertionError: assert 期望 ab 是集合 {'bb', 'aa'} 中的一个元素,实际集合 {'bb', 'aa'} 中没有 ab 元素
FAILED test_demo.py::test_13 - AssertionError: assert 期望 gender 是字典 {'name': '张三丰', 'age': 100} 中的一个 key,实际字典 {'name': '张三丰', 'age': 100} 中没有值...
FAILED test_demo.py::test_14 - assert 期望 aa 不是 bbaa 的子串,实际 aa 是 bbaa 的子串
FAILED test_demo.py::test_16 - AssertionError: assert 期望 aa 不是集合 ['aa', 'bb'] 中的一个元素,实际集合 ['aa', 'bb'] 中有 aa 元素
FAILED test_demo.py::test_17 - AssertionError: assert 期望 aa 不是集合 ('aa', 'bb') 中的一个元素,实际集合 ('aa', 'bb') 中有 aa 元素
FAILED test_demo.py::test_18 - AssertionError: assert 期望 aa 不是集合 {'bb', 'aa'} 中的一个元素,实际集合 {'bb', 'aa'} 中有 aa 元素
FAILED test_demo.py::test_22 - AssertionError: assert 期望 name 不是字典 {'name': '张三丰', 'age': 100} 中的一个 key,实际字典{'name': '张三丰', 'age': 100}中有值为 na...
========================= 12 failed, 11 passed in 0.23s =========================
```

第 4 章　mark 标签的用法

4.1　skip 和 skipif 的使用方法

在 Pytest 中，markers 是用来给测试脚本增加标记的，Pytest 提供了一些内置的标签，下面针对常用的几个标签作详细的讲解。

4.1.1　skip 的用法

skip 标签主要用于跳过执行操作，比如针对一个功能自动化测试的脚本已经实现了，但是产品中此功能尚未实现，如果执行此用例，则没有什么意义，此时就可以通过 skip 将当前脚本标记为跳过，下面 skip 针对测试函数的用法中，括号中的 reason 可以填写原因。

```python
import pytest

@pytest.mark.skip(reason="功能尚未实现,待实现后再运行")
def test_func():
    assert1 == 1
```

从执行结果中可以发现，Pytest 并未执行当前脚本，而是把当前脚本标记为 skip，并且在报告中显示为 skip 状态。

```
(demo-HCIhXOHq) E:\demo> pytest
============================ test session starts ============================
platform win32 -- Python 3.7.9, pytest-7.1.3, pluggy-1.0.0
rootdir: E:\demo
plugins: assume-2.4.3, rerunfailures-10.2
collected 1 item

test_demo.py s                                                        [100%]

============================ 1 skipped in 0.01s ============================

(demo-HCIhXOHq) E:\demo>
```

在测试函数中使用装饰器跳过的时候，整个测试函数都不会执行。在某一些情况下,测试函数内容比较多,希望在执行的过程中检测一些条件,如果条件不满足则跳过后面的测试代码,此时需要在测试函数中使用 skip 跳过测试函数的部分代码,如果当前系统是 Windows 系统,则停止执行,并跳过后面的测试代码,将当前脚本标记为 skip 状态。

```
import pytest
import platform

def test_func():
    print("begin to test...")
    if platform.system() == "Windows":
        pytest.skip("不支持 windows")
    print("end of test")
```

在 Windows 上执行结果中,我们可以发现在判断条件之前的打印是执行了,当条件判断之后就停止执行,并跳过后面的测试代码,这时,条件判断之后的打印未执行,并且测试报告中将当前脚本标记为 skip 状态。

```
(demo-HCIhX0Hq) E:\demo> pytest
============================ test session starts ============================
platform win32 -- Python 3.7.9, pytest-7.1.3, pluggy-1.0.0
rootdir: E:\demo
plugins: assume-2.4.3, rerunfailures-10.2
collected 1 item

test_demo.py s                                                        [100%]

============================ 1 skipped in 0.01s ============================

(demo-HCIhX0Hq) E:\demo>
```

此外,在某些场景下,比如想在一定条件下,整个测试文件中的测试脚本全部跳过不执行,此时就可以采用直接在测试文件中使用 Pytest.skip 进行标记。

```
import sys
import pytest

if sys.platform.startswith("win"):
    pytest.skip("不支持 windows", allow_module_level=True)

def test_func1():
    assert1 == 1

def test_func2():
```

```
        assert1 == 1

    def test_func3():
        assert1 == 1
```

执行结果中看出此时并不会识别当前用例中有多少个用例,而是直接跳过,在测试报告中也只是标记为 skip 状态。

```
(demo - HCIhX0Hq) E:\demo> pytest
=========================== test session starts ===========================
platform win32 -- Python 3.7.9, pytest - 7.1.3, pluggy - 1.0.0
rootdir: E:\demo
plugins: assume - 2.4.3, rerunfailures - 10.2
collected 0 items / 1 skipped

=========================== 1 skipped in 0.01s ===========================

(demo - HCIhX0Hq) E:\demo>
```

如果希望将一个类中的所有测试方法全部跳过,此时只需要在类上使用装饰进行 skip 标记即可,如下所示。

```
import pytest

@pytest.mark.skip
class TestDemo:
    def test_func1(self):
        assert1 == 1

    def test_func2(self):
        assert1 == 1
```

执行结果中发现,此时会识别有多少个用例未执行。

```
(demo - HCIhX0Hq) E:\demo> pytest
=========================== test session starts ===========================
platform win32 -- Python 3.7.9, pytest - 7.1.3, pluggy - 1.0.0
rootdir: E:\demo
plugins: assume - 2.4.3, rerunfailures - 10.2
collected 2 items

test_demo.py ss                                              [100%]

=========================== 2 skipped in 0.01s ===========================
```

```
(demo - HCIhX0Hq) E:\demo>
```

4.1.2 skipif 的用法

skipif 功能跟 skip 类似,都是对测试脚本做 skip 标记,且不执行。不同点是 skipif 可以直接设置跳过条件,当条件为真时,即直接在第一个参数中设置跳过条件。

```
import pytest
import platform

@pytest.mark.skipif(platform.system() == "Windows", reason = "不支持window平台")
def test_func():
    assert1 == 1
```

如下所示,此时因为在 Windows 平台执行,所以 skipif 中的条件为真,即直接跳过不执行。

```
(demo - HCIhX0Hq) E:\demo> pytest
============================ test session starts =========================
platform win32 -- Python 3.7.9, pytest - 7.1.3, pluggy - 1.0.0
rootdir: E:\demo
plugins: assume - 2.4.3, rerunfailures - 10.2
collected 1 item

test_demo.py s                                          [100%]

============================ 1 skipped in 0.01s =========================

(demo - HCIhX0Hq) E:\demo>
```

同样,对于跳过类中所有的测试方法,可直接在类上使用装饰器,即直接在类上加上 skipif 的装饰器。

```
import pytest
import platform

@pytest.mark.skipif(platform.system() == "Windows", reason = "不支持window平台")
class TestDemo():
    def test_01(self):
        assert1 == 1

def test_02(self):
    assert 1 == 1
```

此时,测试类中的两个测试方法均直接跳过,不再执行。

```
(demo-HCIhXOHq) E:\demo> pytest
=========================== test session starts ===========================
platform win32 -- Python 3.7.9, pytest-7.1.3, pluggy-1.0.0
rootdir: E:\demo
plugins: assume-2.4.3, rerunfailures-10.2
collected 2 items

test_demo.py ss                                                      [100%]

============================ 2 skipped in 0.02s ===========================

(demo-HCIhXOHq) E:\demo>
```

当将一个测试文件中所有测试函数和测试方法都跳过时,只需要按照如下的方式使用即可。

```
import pytest
import platform


pytestmark = pytest.mark.skipif(platform.system() == "Windows", reason="不支持window平台")


def test_01():
    assert 1 == 1


def test_02():
    assert 2 == 2
```

如下所示,此时执行结果直接跳过整个测试文件中的所有测试函数。

```
(demo-HCIhXOHq) E:\demo> pytest
=========================== test session starts ===========================
platform win32 -- Python 3.7.9, pytest-7.1.3, pluggy-1.0.0
rootdir: E:\demo
plugins: assume-2.4.3, rerunfailures-10.2
collected 2 items

test_demo.py ss                                                      [100%]

============================ 2 skipped in 0.02s ===========================

(demo-HCIhXOHq) E:\demo>
```

4.2 xfail 和 xpass 的用法

4.2.1 xfail 标记测试脚本

当 xfail 用于标记此用例时可能会失败，脚本失败时，测试报告不会打印错误追踪，只是会显示 xfail 状态。xfail 的主要作用是：当进行测试提前时，产品的某功能尚未开发完成，即可进行自动化脚本开发，此时也可以把这些脚本注释起来，但这不是 Pytest 推荐的做法，Pytest 推荐使用 xfail 标记，虽然产品功能尚未开发完成，但是自动化脚本已经可以跑起来了，只不过在测试报告中会显示 xfail 而已，如下所示。

```
import pytest

@pytest.mark.xfail
def test_01():
    assert 1 == 2
```

我们看执行结果，此时，因为测试脚本中使用了 assert 1==2 模拟断言失败，故执行结果显示测试脚本的执行状态为 xfailed。

```
(demo-HCIhX0Hq) E:\demo> pytest
=========================== test session starts ===========================
platform win32 -- Python 3.7.9, pytest-7.1.3, pluggy-1.0.0
rootdir: E:\demo
plugins: assume-2.4.3, rerunfailures-10.2
collected 1 item

test_demo.py x                                                    [100%]

=========================== 1 xfailed in 0.06s ===========================

(demo-HCIhX0Hq) E:\demo>
```

当产品功能开发完成后，脚本执行的结果会显示 xpass 状态，xpass 不是内置标签，不能用于在脚本中标记，xpass 是指用 xfail 标记的脚本在执行的过程中没有报错，因而脚本的执行状态为 xpass 状态，如下所示。

```
import pytest

@pytest.mark.xfail
def test_01():
    assert1 == 1
```

此时因为断言成功了,所以脚本执行状态为 xpassed。

```
(demo-HCIhXOHq) E:\demo> pytest
========================== test session starts ==========================
platform win32 -- Python 3.7.9, pytest-7.1.3, pluggy-1.0.0
rootdir: E:\demo
plugins: assume-2.4.3, rerunfailures-10.2
collected 1 item

test_demo.py X                                               [100%]

========================== 1 xpassed in 0.01s ==========================

(demo-HCIhXOHq) E:\demo>
```

那么正常情况下,说明此时产品功能已经够用了,在实际应用中,这种情况下就可以将 xfail 的标记去掉了,当然不想去掉也没问题,在本章后面的内容中我们会针对这种场景给出对应的解决方案。

4.2.2　xfail 根据条件判断标记测试脚本

在一些场景下,需要通过判断一些条件来标记测试脚本,当满足一定条件时,就可以确认是否会失败,即此时可以将满足一定条件时标记为 xfail,之后标记后的代码则不会被执行。当操作系统为 Windows 系统时则标记为失败。

```python
import pytest
import platform

def test_01():
    if platform.system() == "Windows":
        pytest.xfail("Windows系统暂不支持")
        print("测试,windows系统下不应该打印出当前信息")
    print("不会打印当前信息")
```

上述执行结果如下所示,-s 参数可以将 print 的信息打印出来,在 pytest.xfail 标记后的 print 语句均未打印信息,即在测试函数中,使用 xfail 标记后的代码均不会执行。

```
(demo-HCIhXOHq) E:\demo> pytest -s
========================== test session starts ==========================
platform win32 -- Python 3.7.9, pytest-7.1.3, pluggy-1.0.0
rootdir: E:\demo
plugins: assume-2.4.3, rerunfailures-10.2
collected 1 item
```

```
test_demo.py x
```

```
========================= 1 xfailed in 0.06s =========================
```

```
(demo - HCIhXOHq) E:\demo >
```

4.2.3 动态启用 xfail 标记

在@pytest.mark.xfail 装饰器的参数中,也可以添加条件判断,这里需要特别注意,不是根据一定的条件判断是否为 xfail 或者 xpass,而是根据 xfail 标记的含义去判断。如根据代码中@pytest.mark.xfail 后的条件判断时,如果在 Windows 上执行,即条件为真,含义是启动 xfail 标记。最终状态是 xfail 还是 xpass,也是由脚本中的断言语句决定的。

```
import pytest
import platform

@pytest.mark.xfail(platform.system() == "Windows", reason = "不支持 windows 系统")
def test_01():
    assert 1 == 1
```

如下面代码所示,在 Windows 系统上执行时,这里 @pytest.mark.xfail 装饰器中条件为真,表示对 test_01 测试函数启用 xfail 标记,因 test_01 测试函数中的断言为真,所有执行结果的状态为 xpassed。

```
(demo - HCIhXOHq) E:\demo > pytest
========================= test session starts =========================
platform win32 -- Python 3.7.9, pytest - 7.1.3, pluggy - 1.0.0
rootdir: E:\demo
plugins: assume - 2.4.3, rerunfailures - 10.2
collected 1 item

test_demo.py X                                              [100%]

========================= 1 xpassed in 0.01s =========================

(demo - HCIhXOHq) E:\demo >
```

下面将测试代码修改如下,即在 Windows 系统运行时,@pytest.mark.xfail 后面的条件此时为假,含义为对当前用例不启用 xfail 标记,只是一个普通用例。

```
import pytest
import platform
```

```python
@pytest.mark.xfail(platform.system() == "Linux", reason = "不支持 windows 系统")
def test_01():
    assert 1 == 1
```

执行结果如下所示,此时为关闭 xfail 标记为一个普通的测试脚本,所以,执行的结果为 passed 状态,而不是 xpassed 状态。

```
(demo-HCIhXOHq) E:\demo> pytest
========================== test session starts ==========================
platform win32 -- Python 3.7.9, pytest-7.1.3, pluggy-1.0.0
rootdir: E:\demo
plugins: assume-2.4.3, rerunfailures-10.2
collected 1 item

test_demo.py .                                                     [100%]

=========================== 1 passed in 0.01s ===========================

(demo-HCIhXOHq) E:\demo>
```

下面继续将测试代码修改,让断言错误。

```python
import pytest
import platform


@pytest.mark.xfail(platform.system() == "Linux", reason = "不支持 windows 系统")
def test_01():
    assert 1 == 2
```

执行结果如下所示,此时 xfail 标记为关闭状态,因为断言错误所以整个脚本的执行为失败状态,跟普通的测试脚本是完全一样的。

```
(demo-HCIhXOHq) E:\demo> pytest
========================== test session starts ==========================
platform win32 -- Python 3.7.9, pytest-7.1.3, pluggy-1.0.0
rootdir: E:\demo
plugins: assume-2.4.3, rerunfailures-10.2
collected 1 item

test_demo.py F                                                     [100%]

================================ FAILURES ================================
_____ test_01 _____

    @pytest.mark.xfail(platform.system() == "Linux", reason = "不支持 windows 系统")
```

```
    def test_01():
>       assert 1 == 2
E       assert 1 == 2

test_demo.py:6: AssertionError
================= short test summary info =================
FAILED test_demo.py::test_01 - assert 1 == 2
==================== 1 failed in 0.06s ====================

(demo-HCIhXOHq) E:\demo>
```

4.2.4 @pytest.mark.xfail 只设置 reason 参数

当 @pytest.mark.xfail 装饰器只设置 reason 参数时，表示当前用例启用的是 xfail 标记，reason 只是用来说明理由的，用例执行结果只能是 xpass 或者 xfail。如下所示。

```
import pytest


@pytest.mark.xfail(reason = "功能尚未开发完成")
def test_01():
    assert 1 == 2
```

上述执行结果显示，此时和不加任何参数的时候结果是一样的，reason 参数主要用来标记当前脚本标记为 xfail 的原因。

```
(demo-HCIhXOHq) E:\demo> pytest
==================== test session starts ====================
platform win32 -- Python 3.7.9, pytest-7.1.3, pluggy-1.0.0
rootdir: E:\demo
plugins: assume-2.4.3, rerunfailures-10.2
collected 1 item

test_demo.py x                                       [100%]

================== 1 xfailed in 0.06s ====================

(demo-HCIhXOHq) E:\demo>
```

4.2.5 @pytest.mark.xfail 通过 run 参数设置是否执行

@pytest.mark.xfail 装饰器可以通过设置 run 参数来限定是否执行测试函数，默认情况下，不指定 run 参数时即执行测试函数内容，当设置 run 的参数值为 False 时，则会直接将用例标记为 xfail 状态，而不去执行测试脚本，test_demo.py 代码如下所示。

```python
import pytest

@pytest.mark.xfail(run=False)
def test_01():
    print("\n in test_01...")

@pytest.mark.xfail(run=True)
def test_02():
    print("\n in test_02...")

@pytest.mark.xfail()
def test_03():
    print("\n in test_03...")
```

可以看出，此时三个测试函数中都没有断言语句，按照断言的逻辑，三个测试函数都应该断言成功，而其中 test_01 中 run 参数设置为 False，即此时 test_01 不会执行测试函数，而是直接标为 xfail，结果如下所示。这时并未打印"in test_01"，而且有一个用例状态为 xfail 状态。

```
(demo-HCIhX0Hq) E:\demo> pytest
==================== test session starts ====================
platform win32 -- Python 3.7.9, pytest-7.1.3, pluggy-1.0.0
rootdir: E:\demo
plugins: assume-2.4.3, rerunfailures-10.2
collected 3 items

test_demo.py xXX                                        [100%]

================ 1 xfailed, 2 xpassed in 0.09s ================

(demo-HCIhX0Hq) E:\demo>
```

4.2.6 xpassed 用例显示为失败

xfail 本意是预料到或期望用例是失败的，但当脚本断言成功后，脚本执行结果会变为 xpass，说明通过了，与原本期望的 xfail 不一致了。因此在某些场景下，我们希望 xpass 状态的用例显示为失败，而 xfail 的用例依然为 xfail 状态，此时只需要使用 strict 参数执行，将 strict 参数设置为 True，则执行结果原本为 xpass 的用例此时将变为失败。

```python
import pytest
```

```
@pytest.mark.xfail(strict = True)
def test_01():
    assert 1 == 1

@pytest.mark.xfail(strict = True)
def test_02():
    assert 1 == 2
```

执行结果如下面代码所示，此时 test_01 中原本执行结果为 xpass 状态变为 FAILED 状态了。

```
(demo - HCIhX0Hq) E:\demo> pytest
================== test session starts ==================
platform win32 -- Python 3.7.9, pytest - 7.1.3, pluggy - 1.0.0
rootdir: E:\demo
plugins: assume - 2.4.3, rerunfailures - 10.2
collected 2 items

test_demo.py Fx                                   [100%]


======================= FAILURES ========================
_____ test_01 _____
[XPASS(strict)]
================= short test summary info ===============
FAILED test_demo.py::test_01
================ 1 failed, 1 xfailed in 0.06s ===========

(demo - HCIhX0Hq) E:\demo>
```

4.2.7 使 xfail 标记失效的方法

前面提过，xfail 的主要作用是在功能尚未支持的时候，暂时将脚本标记为 xfail，到了后期功能支持了，才放开这些标记。当然也可以通过修改脚本，将这些标记装饰器代码注释掉或者删除。此外，Pytest 还提供了一种更为便捷的方式，即不需要注释或者删除原有的标记代码就可执行，只需要在 Pytest 命令中加上 --runxfail 参数就能实现演示代码如下所示。

```
import pytest

@pytest.mark.xfail(run = True)
def test_01():
    print("\n in test_01...")
```

```
@pytest.mark.xfailj(run = False)
def test_02():
    print("\n in test_02...")


@pytest.mark.xfail
def test_03():
    print("\n in test_03...")
```

按照前面 xfail 的分析,test_02 不会执行,且会有两个被标记为 xpassed,其中一个被标记为 xfail,这里加上 -- runxfail 参数,但此时好像没有 xfail 标记一样,而是完全按照普通的测试脚本执行了。这就解决了一个重要的场景问题,即当产品功能尚未开发完成时,我们开发了许多自动化脚本,然后将这些脚本标记为 xfail,而当产品功能开发完成后,不想将开发的脚本删除或者注释 xfail 的标记,就可以通过 -- runxfail 参数的方式直接将所有的脚本执行了,这就是这个参数的实际应用价值。

```
(demo - HCIhXOHq) E:\demo> pytest -- runxfail
==================== test session starts ====================
platform win32 -- Python 3.7.9, pytest - 7.1.3, pluggy - 1.0.0
rootdir: E:\demo
plugins: assume - 2.4.3, rerunfailures - 10.2
collected 3 items

test_demo.py ...                                       [100%]

==================== warnings summary ====================
test_demo.py:7
    E:\demo\test_demo.py:7: PytestUnknownMarkWarning: Unknown pytest.mark.xfailj - is this a typo?  You can register custom marks to avoid this warning - for details, see
    https://docs.pytest.org/en/stable/how-to/mark.html
        @pytest.mark.xfailj(run = False)

-- Docs: https://docs.pytest.org/en/stable/how-to/capture-warnings.html
============== 3 passed, 1 warning in 0.02s ===============

(demo - HCIhXOHq) E:\demo>
```

4.3 importorskip 的用法

在写自动化脚本的时候,对于导入包作为被测对象导入时失败,或者按照正常导入包逻辑,开发出的自动化脚本无法运行情况,Pytest 提供了 importorskip 内置标签,可在导入包失败时,自动将当前文件中的测试脚本标记为 skipped 状态。

如情况导入 numpy 库时，因为当前的环境尚未安装 numpy 库，可能会导入失败，此时 Pytest 可自动将当前文件的测试函数全部标记为 skipped。

```
import pytest

np = pytest.importorskip("numpy")

def test_01():
    print(dir(np))
    assert 1 == 1
```

执行结果如下所示。

```
(demo-HCIhXOHq) E:\demo> pytest -s
=================== test session starts ===================
platform win32 -- Python 3.7.9, pytest-7.1.3, pluggy-1.0.0
rootdir: E:\demo
plugins: assume-2.4.3, rerunfailures-10.2
collected 0 items / 1 skipped

=================== 1 skipped in 0.01s ===================

(demo-HCIhXOHq) E:\demo>
```

当模块存在且导入成功时，返回值即是导入的模块对象，换言之，可以直接将返回值赋值给一个对象。在接下来的测试脚本中，我们直接使用调用导入模块的方法，编写测试代码，这里将模块修改为导入 math 库。

```
import pytest

math = pytest.importorskip("math")

def test_01():
    print(dir(math))
    assert 1 == 1
```

执行结果显示导入成功，math 变量即导入的 math 库。通常在测试函数中，打印出的都是 math 库的方法或属性。

```
(demo-HCIhXOHq) E:\demo> pytest -s
=================== test session starts ===================
platform win32 -- Python 3.7.9, pytest-7.1.3, pluggy-1.0.0
rootdir: E:\demo
plugins: assume-2.4.3, rerunfailures-10.2
collected 1 item
```

```
test_demo.py ['__doc__', '__loader__', '__name__', '__package__', '__spec__', 'acos',
'acosh', 'asin', 'asinh', 'atan', 'atan2', 'atanh', 'ceil', 'copysign', 'cos', 'cosh', 'degrees', 'e',
'erf', 'erfc', 'exp', 'expm1', 'fabs', 'factorial', 'floor', 'fmod', 'frexp', 'fsum', 'gamma', 'gcd',
'hypot', 'inf', 'isclose', 'isfinite', 'isinf', 'isnan', 'ldexp', 'lgamma', 'log', 'log10', 'log1p',
'log2', 'modf', 'nan', 'pi', 'pow', 'radians', 'remainder', 'sin', 'sinh', 'sqrt', 'tan', 'tanh', 'tau',
'trunc']

===================== 1 passed in 0.01s =====================

(demo-HCIhX0Hq) E:\demo>
```

4.4 注册并使用自定义 mark 标签

4.4.1 直接使用自定义 mark 标签

在 Pytest 中，可以通过@pytest.mark.xxx 的方式直接自定义标签使用，比如欲对一个测试函数增加 smoke 标签，直接在测试函数中使用@pytest.mark.smoke 即可。

```
import pytest

@pytest.mark.smoke
def test_01():
    assert 1 == 1
```

测试函数增加自定义标签的作用是对自动化脚本进行分类标记，如测试活动常常分为冒烟测试、功能测试、系统测试、性能测试，测试用例同样也分为冒烟测试用例、功能测试用例、系统测试用例、性能测试用例，因此，自动化脚本就是同步化自动化脚本的分类。Pytest 中的 mark 主要就是实现这种功能的。在正式使用 mark 对自动化脚本标记之前，我们先看下上面这个自动化脚本的执行结果。这里执行成功了，但是可以看到一条告警信息，从告警信息可以看出，此时 smoke 标签为未识别标签，即这里没有注册，而是直接使用。

```
(demo-HCIhX0Hq) E:\demo> pytest
===================== test session starts =====================
platform win32 -- Python 3.7.9, pytest-7.1.3, pluggy-1.0.0
rootdir: E:\demo
plugins: assume-2.4.3, rerunfailures-10.2
collected 1 item
```

```
test_demo.py .                                                    [100%]

===================== warnings summary =====================
test_demo.py:3
  E:\demo\test_demo.py:3: PytestUnknownMarkWarning: Unknown pytest.mark.smoke - is this a typo?
  You can register custom marks to avoid this warning - for details, see https://docs.pytest.org/en/stable/how-to/mark.html
    @pytest.mark.smoke

-- Docs: https://docs.pytest.org/en/stable/how-to/capture-warnings.html
===================== 1 passed, 1 warning in 0.02s =================

(demo-HCIhXOHq) E:\demo>
```

下面再写一个测试函数,通过 mark 直接定义 smoke 和 function 两个标签,然后验证在不注册的情况下,通过 Pytest 命令是否可以做到挑选用例执行。

```
import pytest

@pytest.mark.smoke
def test_01():
    assert 1 == 1


@pytest.mark.function
def test_02():
    assert 1 == 1
```

我们看到,Pytest 命令可以通过-m 参数指定标签挑选脚本来执行,如通过 Pytest-m smoke 命令可以挑选打了 smoke 标签的脚本,可以发现,此时虽然未对 smoke 和 function 标签进行注册,但仍然可以通过指定标签的方式挑选测试脚本。不注册而直接使用标签的一个缺点就是会显示告警信息。后面将继续展示通过注册标签然后再使用的情形。

```
(demo-HCIhXOHq) E:\demo> pytest -m smoke
===================== test session starts =====================
platform win32 -- Python 3.7.9, pytest-7.1.3, pluggy-1.0.0
rootdir: E:\demo
plugins: assume-2.4.3, rerunfailures-10.2
collected 2 items / 1 deselected / 1 selected

test_demo.py .                                                    [100%]
```

```
==================== warnings summary ====================
test_demo.py:3
    E:\demo\test_demo.py:3: PytestUnknownMarkWarning: Unknown pytest.mark.smoke - is
this a typo? You can register custom marks to avoid this warning - for details, see
    https://docs.pytest.org/en/stable/how-to/mark.html
        @pytest.mark.smoke

test_demo.py:8
    E:\demo\test_demo.py:8: PytestUnknownMarkWarning: Unknown pytest.mark.function -
is this a typo? You can register custom marks to avoid this warning - for details, see
    https://docs.pytest.org/en/stable/how-to/mark.html
        @pytest.mark.function

-- Docs: https://docs.pytest.org/en/stable/how-to/capture-warnings.html
======= 1 passed, 1 deselected, 2 warnings in 0.02s =======

(demo-HCIhXOHq) E:\demo>
```

4.4.2 通过 conftest.py 文件重写 pytest_configure 函数的注册标签

利用在测试脚本根目录下创建 conftest.py 文件,再在 conftest.py 文件中重写 pytest_configure 函数的方式可以注册标签。如注册 smoke 标签和 function 标签,函数名是固定的,即 pytest_configure,参数也是固定的,即 config,唯一需要变更的就是注册的标签的名字和描述。

```
def pytest_configure(config):
    config.addinivalue_line(
        "markers", "smoke: smoke test"
    )
    config.addinivalue_line(
        "markers", "function: system test"
    )
```

我们在对 Pytest_configure 函数重写后,再编写测试函数,也就是 test_01 和 test_02,并使用 smoke 和 function 进行标记。

```
import pytest


@pytest.mark.smoke
def test_01():
    assert 1 == 1


@pytest.mark.function
```

```
def test_02():
    assert 1 == 1
```

此时脚本的执行结果如下所示,即告警消除了,同时挑选 smoke 标签的脚本也执行了。在企业应用中,倘若在执行脚本的时候发现有大量提示未知标签的告警,那么说明未对告警进行注册。

```
(demo-HCIhXOHq) E:\demo> pytest -m smoke
==================== test session starts ====================
platform win32 -- Python 3.7.9, pytest-7.1.3, pluggy-1.0.0
rootdir: E:\demo
plugins: assume-2.4.3, rerunfailures-10.2
collected 2 items / 1 deselected / 1 selected

test_demo.py .                                        [100%]

============= 1 passed, 1 deselected in 0.01s =============

(demo-HCIhXOHq) E:\demo>
```

4.4.3 通过 Pytest.ini 文件配置注册标签

除了在 conftest.py 文件中重写 Pytest_configure 函数,Pytest 还提供了一种更加便捷的注册标签的方式,即在 Pytest.ini 配置文件中进行配置,配置的语法相对比较简洁,即在测试脚本的根目录下创建 Pytest.ini 文件,比如同样注册 smoke 和 function 标签,只需要在 Pytest.ini 中编写做配置即可。

```
[pytest]
markers =
    smoke: smoke tests
    function: function tests
```

此时,继续使用测试函数来验证一下,并分别为 test_01 和 test_02,使用 smoke 和 function 打标签。

```
import pytest


@pytest.mark.smoke
def test_01():
    assert 1 == 1


@pytest.mark.function
def test_02():
```

```
    assert 1 == 1
```

我们通过 Pytest - m smoke 挑选打了 smoke 标签的测试函数来执行,可以发现,此时同样不会出现打印告警,说明注册标签生效了。

```
(demo - HCIhX0Hq) E:\demo> pytest - m smoke
==================== test session starts ====================
platform win32 -- Python 3.7.9, pytest - 7.1.3, pluggy - 1.0.0
rootdir: E:\demo, configfile: pytest.ini
plugins: assume - 2.4.3, rerunfailures - 10.2
collected 2 items / 1 deselected / 1 selected

test_demo.py .                                         [100%]

============= 1 passed, 1 deselected in 0.01s =============

(demo - HCIhX0Hq) E:\demo>
```

很容易发现,通过 Pytest.ini 配置文件配置注册标签的方式更加便捷简单,因此在实际自动化脚本开发中,也推荐大家通过 Pytest.ini 配置文件的方式配置注册使用标签。

4.4.4 通过标签灵活挑选测试脚本执行

Mark 标签在自动化测试中非常重要,下面详细讲解在实际应用中如何综合使用 mark 标签。我们先通过 Pytest.ini 注册三个标签,smoke、function 和 performance,分别表示冒烟测试脚本、功能测试脚本和性能测试脚本,Pytest.ini 配置文件中注册标签的配置如下所示:

```
[pytest]
markers =
    smoke: smoke tests
    function: function tests
    performance: performane tests
```

我们来编写测试函数脚本,这里编写了三个测试函数,test_01、test_02 和 test_03。注意,针对每个测试函数,是可以打多个标签的,这是合理的。比如在实际开发中,某个用例既可以作为功能测试用例,又可以作为冒烟测试用例。实际上冒烟测试用例就是从功能测试用例中挑出部分最基本的用例。在下面的代码中,为了更好演示标签的使用,对每个测试函数均加了两个标签。

```
import pytest


@pytest.mark.function
```

```python
@pytest.mark.smoke
def test_01():
    assert 1 == 1


@pytest.mark.function
@pytest.mark.performance
def test_02():
    assert 1 == 1

@pytest.mark.performance
@pytest.mark.smoke
def test_03():
    assert 1 == 1
```

这里通过指定 smoke 标签发现此时执行了两个测试用例，由于 test_01 和 test_03 均加了 smoke 标签，因此当指定 smoke 标签的时候，会将所有打了 smoke 标签的脚本做执行。

```
(demo-HCIhXOHq) E:\demo> pytest -m smoke
=================== test session starts ===================
platform win32 -- Python 3.7.9, pytest-7.1.3, pluggy-1.0.0
rootdir: E:\demo, configfile: pytest.ini
plugins: assume-2.4.3, rerunfailures-10.2
collected 3 items / 1 deselected / 2 selected

test_demo.py ..                                    [100%]

============= 2 passed, 1 deselected in 0.02s =============

(demo-HCIhXOHq) E:\demo>
```

另外，Pytest 还可以通过逻辑关系词 and，or 和 not 类指定满足一定逻辑关系的标签组合挑选测试脚本来执行。and 表示同时满足，or 表示逻辑或，相当于并集，not 相当于非，此外当使用逻辑关系词时，需要将指定的标签逻辑关系部分使用引号括起来，比如想执行同时打了 smoke 和 function 的标签，使用的命令就是 Pytest -m "smoke and function"，此时只执行了一个标签，因为同时打了 smoke 和 function 标签的脚本只有 test_01。

```
(demo-HCIhXOHq) E:\demo> pytest -m "smoke and function"
=================== test session starts ===================
platform win32 -- Python 3.7.9, pytest-7.1.3, pluggy-1.0.0
rootdir: E:\demo, configfile: pytest.ini
```

```
plugins: assume-2.4.3, rerunfailures-10.2
collected 3 items / 2 deselected / 1 selected

test_demo.py .                                              [100%]

=============1 passed, 2 deselected in 0.01s =============

(demo-HCIhXOHq) E:\demo>
```

当想执行打了 smoke 或者 function 标签的脚本时,即使用逻辑关系词 or,使用的命令为 pytest -m "smoke or function",因为此时三个测试函数要么打了 smoke 标签,要么打了 function 标签,因此这里显示执行了三个测试函数。

```
(demo-HCIhXOHq) E:\demo> pytest -m "smoke or function"
===================test session starts ===================
platform win32 -- Python 3.7.9, pytest-7.1.3, pluggy-1.0.0
rootdir: E:\demo, configfile: pytest.ini
plugins: assume-2.4.3, rerunfailures-10.2
collected 3 items

test_demo.py ...                                            [100%]

===================3 passed in 0.02s =====================

(demo-HCIhXOHq) E:\demo>
```

同理,如果加入想执行的除性能测试以外的所有用例,即执行除了打 performance 标签以外的所有用例,就需要使用 not 逻辑词,执行的命令为 Pytest -m "not performance"。由于 test_02 和 test_03 均打了 performance 标签,因此这里只执行了一个测试脚本。

```
(demo-HCIhXOHq) E:\demo> pytest -m "not performance"
===================test session starts ===================
platform win32 -- Python 3.7.9, pytest-7.1.3, pluggy-1.0.0
rootdir: E:\demo, configfile: pytest.ini
plugins: assume-2.4.3, rerunfailures-10.2
collected 3 items / 2 deselected / 1 selected

test_demo.py .                                              [100%]

=============1 passed, 2 deselected in 0.02s =============

(demo-HCIhXOHq) E:\demo>
```

第 5 章　Pytest 测试用例的执行策略

如使 Pytest 框架支持不同场景下各种执行策略的需求,可以通过命令行参数或安装一些第三方插件完成,本章将围绕执行脚本策略详细展开讲解。

5.1　在遇到用例失败时如何停止执行

Pytest 命令在默认的情况下会执行所有的测试脚本,其中 test_01 中断言将失败,test_02 中断言会成功。

```
def test_01():
    assert 1 == 2

def test_02():
    assert 1 == 1
```

当直接使用 Pytest 命令时,会默认将 test_01 和 test_02 两个测试脚本全部执行,test_01 和 test_02 将全部执行。

```
(demo-HCIhXOHq) E:\demo> pytest
===================== test session starts =====================
platform win32 -- Python 3.7.9, pytest-7.1.3, pluggy-1.0.0
rootdir: E:\demo
plugins: assume-2.4.3, rerunfailures-10.2
collected 2 items

test_demo.py F.                                          [100%]

============================= FAILURES ============================
_____ test_01 _____

    def test_01():
>       assert 1 == 2
E       assert 1 == 2

test_demo.py:2: AssertionError
================== short test summary info ==================
```

```
FAILED test_demo.py::test_01 - assert 1 == 2
================1 failed, 1 passed in 0.06s ================

(demo-HCIhX0Hq) E:\demo>
```

当我们在开发自动化脚本时，并不希望一次性将所有脚本执行全部完成，而是希望当遇到用例失败时能停下来，这样便于随时进行问题定位和解决，Pytest 支持这样的执行策略，只需要使用-x 参数即可。使用-x 参数时，执行测试脚本 test_01 断言失败后就会停下来，不再执行后面的测试用例，这种场景对于在开发脚本或者调试自动化脚本的时候是非常有用的。

```
(demo-HCIhX0Hq) E:\demo> pytest -x
==================== test session starts ====================
platform win32 -- Python 3.7.9, pytest-7.1.3, pluggy-1.0.0
rootdir: E:\demo
plugins: assume-2.4.3, rerunfailures-10.2
collected 2 items

test_demo.py F

========================== FAILURES ==========================
_____ test_01 _____

    def test_01():
>       assert 1 == 2
E       assert 1 == 2

test_demo.py:2: AssertionError
================= short test summary info =================
FAILED test_demo.py::test_01 - assert 1 == 2
!!!!!!!!!!!!!!!! stopping after 1 failures !!!!!!!!!!!!!!!!
===================== 1 failed in 0.06s =====================

(demo-HCIhX0Hq) E:\demo>
```

5.2　如何在用例失败时打印局部变量

使用 Pytest 命令执行自动化脚本时，如果断言错误，默认只会显示断言语句的失败信息。如在测试函数 test_01 中定义了多个局部变量，断言失败后很难定位，往往不知道这些变量中是哪个环节出了问题。比较低级一点的做法是在每个变量后面将变量值打印出来,显然这种方式是很麻烦的。

```
def test_01():
    a = 1
    b = a + 1
    c = b + 1
    assert c == 5
```

如果直接使用 Pytest 命令执行的结果,会只看到变量 c=3,与期望值 5 不相等,在具体的 a、b、c 变量之间的计算中间到底哪个环节出了问题,或从哪个变量开始出错,执行结果中是很难看出来的。而如果在执行结果中能将这个测试函数中的所有局部变量都打印出来,那就非常方便。

```
(demo-HCIhXOHq) E:\demo> pytest
=================== test session starts ====================
platform win32 -- Python 3.7.9, pytest-7.1.3, pluggy-1.0.0
rootdir: E:\demo
plugins: assume-2.4.3, rerunfailures-10.2
collected 1 item

test_demo.py F                                       [100%]

========================= FAILURES =========================
_____ test_01 _____

    def test_01():
        a = 1
        b = a + 1
        c = b + 1
>       assert c == 5
E       assert 3 == 5

test_demo.py:5: AssertionError
================= short test summary info =================
FAILED test_demo.py::test_01 - assert 3 == 5
===================== 1 failed in 0.06s ====================

(demo-HCIhXOHq) E:\demo>
```

Pytest 提供了这样的功能;在使用 Pytest 命令的时候使用-l 参数就可以很清晰地看出 a=1,b=2,c=3。在实际应用中,这样可以很容易根据各个变量的实际值判断出到底在哪个环节出现了问题。

```
(demo-HCIhXOHq) E:\demo> pytest -l
```

```
==================== test session starts ====================
platform win32 -- Python 3.7.9, pytest-7.1.3, pluggy-1.0.0
rootdir: E:\demo
plugins: assume-2.4.3, rerunfailures-10.2
collected 1 item

test_demo.py F                                         [100%]

========================= FAILURES =========================
_____ test_01 _____

    def test_01():
        a = 1
        b = a + 1
        c = b + 1
>       assert c == 5
E       assert 3 == 5

a          = 1
b          = 2
c          = 3

test_demo.py:5: AssertionError
================== short test summary info ==================
FAILED test_demo.py::test_01 - assert 3 == 5
==================== 1 failed in 0.06s ====================

(demo-HCIhXOHq) E:\demo>
```

此外,当用例通过后,使用-l参数不会打印局部变量,因此不用担心执行自动化脚本,且用例断言成功时是否会有很多打印。比如将代码稍作修改,把断言内容修改为期望变量 c 等于 3,此时断言显然成功。

```
def test_01():
    a = 1
    b = a + 1
    c = b + 1
    assert c == 3
```

当断言成功时,使用-l参数不会打印局部变量,我们发现 Pytest 的这个功能在执行自动化脚本的时候是非常有用的。

```
(demo-HCIhXOHq) E:\demo> pytest -l
==================== test session starts ====================
```

```
platform win32 -- Python 3.7.9, pytest-7.1.3, pluggy-1.0.0
rootdir: E:\demo
plugins: assume-2.4.3, rerunfailures-10.2
collected 1 item

test_demo.py .                                       [100%]

==================== 1 passed in 0.01s ====================

(demo-HCIhXOHq) E:\demo>
```

5.3 如何在用例执行失败时使用 pdb 进行调试

当在 linux 环境下进行脚本调试或用调试模式进行定位问题时，最好能进入调试模式，Pytest 框架即提供了这样的功能。在 Pytest 框架中，加上-pdb 参数即可在自动化脚本断言报错的时候直接进入调试模式，这比 Python 语言中的使用更简单一些。下面就通过一个实例来演示如何使用 Pytest 中的 pdb 调试模式。

编写自动化调试情况如下所示。

```
def get_abs(x):
    if x>=0:
        return x
    else:
        return -x

def test_01():
    a=-10
    b=get_abs(a)
    assert b==-10
```

这里定义了一个获取绝对值的函数，通过分析得知代码 b 的值为 10，而在测试脚本中断言期望值-10，显然这里断言会报错。下面就通过 Pytest-pdb 命令在断言报错时进入调试模式来一步一步地演示 pdb 调试的方法。

当 Pytest-pdb 进入调试模式，会有（Pdb）提示符出现。

在调试模式下，使用 l 命令（注意是字母 L 的小写，不是数字 1）可以查看当前位置前后的 5 行代码，比如当前位置为断言 assert b==10，l 命令即可以查看在此前后各 5 行的代码，如果后面没有代码了，则只显示 6～11 行的代码。

l 命令还可以通过指定起始和结束的行号来指定显示代码行，比如 l 1,20 表示会显示第 1～20 行的代码，这里因为共 11 行代码，因此只显示了第 1～11 行的代码。

在调试模式下，通过命令 p 可以打印变量的值，如打印了局部变量 a 和 b 的值，就很容易定位问题。pp 也可以打印变量值，跟 p 命令基本一样。

此外，在调试模式中，还可以调用函数，只不过调用函数时对于调用的返回值需要通过 p 打印出来，因为在调试模式下，如果将调用的函数赋值给新的变量没有什么意义。可利用对 get_abs 函数分别传入 5、0、-5 来测试函数返回值的正确性。

退出调试模式可以通过 q 命令来完成。

整个调试过程如下所示。

```
(demo-HCIhX0Hq) E:\demo> pytest --pdb
==================== test session starts ====================
platform win32 -- Python 3.7.9, pytest-7.1.3, pluggy-1.0.0
rootdir: E:\demo
plugins: assume-2.4.3, rerunfailures-10.2
collected 1 item

test_demo.py F
>>>>>>>>>>>>>>>>>>>>>>>> traceback >>>>>>>>>>>>>>>>>>>>>>>>

    def test_01():
        a = -10
        b = get_abs(a)
>       assert b == -10
E       assert 10 == -10

test_demo.py:10: AssertionError
>>>>>>>>>>>>>>>>>>>>>>>> entering PDB >>>>>>>>>>>>>>>>>>>>>>>>

>>>>>>>>>>>>>> PDB post_mortem (IO-capturing turned off) >>>>>>>>>>>>
> e:\demo\test_demo.py(10)test_01()
-> assert b == -10
(Pdb) l
  5             return -x
  6
  7     def test_01():
  8         a = -10
  9         b = get_abs(a)
 10 ->      assert b == -10
 11
[EOF]
(Pdb) l 1,20
  1     def get_abs(x):
  2         if x >= 0:
```

```
  3              return x
  4          else:
  5              return -x
  6
  7   def test_01():
  8       a = -10
  9       b = get_abs(a)
 10 ->    assert b == -10
 11
[EOF]
(Pdb) p a
-10
(Pdb) p b
10
(Pdb) p get_abs(5)
5
(Pdb) p get_abs(0)
0
(Pdb) p get_abs(-5)
5
(Pdb) pp a
-10
(Pdb) pp b
10
(Pdb) q


================ short test summary info =================
FAILED test_demo.py::test_01 - assert 10 == -10
!!!!!!!!! _pytest.outcomes.Exit: Quitting debugger !!!!!!!!!
============== 1 failed in 80.77s (0:01:20) ===============

(demo-HCIhXOHq) E:\demo >
```

5.4 用例失败后如何重新执行

在自动化脚本的执行过程中,存在这样的场景,如会有某一个或者某一类脚本不稳定,即在网络环境、网速快慢等的干扰下,可能会出现失败的情况,但是一般如果脚本失败了再执行一遍就能通过了。从代码开发的角度看,其实在自动化脚本中这些问题都容易解决,但在自动化脚本开发调试完成后,在大批量脚本连跑的时候会陆续发现这种问题，如果全部在自动化脚本中处理妥当，要花一定的时间，且会存在风险。而在后续脚本执行过程中陆陆续续被发现，有时候时间紧任务重，不允许脚本再重跑，因此在这种情况下，我们迫切希望有这样一个功能，即当遇到一个脚本失败时，我们可以让这个脚本继续多执行几次，直到通过位置，当然也可以设置最大重跑次数，比如设置最多重跑5次。如果重跑5次脚本依然失败，那就可以认为这个脚本没有通过，需要后续进一步定位和分析。

我们这里需要借助 Pytest 的第三方插件 Pytest-rerunfailures 来完成。在使用前，需要安装一下这个插件，安装命令如下所示。

```
pip install pytest-rerunfailures
```

安装插件后，可以直接在 Pytest 命令后加上比如-rerun 2，表示如果脚本失败则最多再重新执行2次，如下所示。

```
def test_01():
    print("in test_01")
    assert 1 == 2


def test_02():
    print("in test_02")
    assert 1 == 1
```

观察上述内容，发现有两个测试函数，其中 test_01 断言失败，test_02 断言成功，使用-reruns 2 来执行，-s 参数作用是将打印的信息打印出来。我们发现这里 test_01 总共执行了3次，因为设置的重执行次数为2，重新执行的最后一次仍然断言失败，此时是把 test_01 标记为失败了，而 test_02 断言成功，因此只执行了一次。

```
(demo-HCIhXOHq) E:\demo> pytest -s --reruns 2
==================== test session starts ====================
platform win32 -- Python 3.7.9, pytest-7.1.3, pluggy-1.0.0
rootdir: E:\demo
plugins: assume-2.4.3, rerunfailures-10.2
collected 2 items

test_demo.py in test_01
Rin test_01
Rin test_01
Fin test_02



========================= FAILURES =========================
_____ test_01 _____
```

```
    def test_01():
        print("in test_01")
>       assert 1 == 2
E       assert 1 == 2

test_demo.py:3: AssertionError
================ short test summary info =================
FAILED test_demo.py::test_01 - assert 1 == 2
========== 1 failed, 1 passed, 2 rerun in 0.07s ==========
```

(demo-HCIhXOHq) E:\demo >

此外，失败重执行插件还可以设置在遇到失败后重执行过程中每次的时间间隔，比如--reruns-dellay 3可以设置每次重新执行的时间间隔为3 s，但从执行结果看不出效果。在执行的过程中确实能看到当执行失败后有等待，另外，通过最后耗时6.07 s也能看出来，前面的执行结果是耗时0.07 s，显然这是等待了两个3 s的执行结果。

```
(demo-HCIhXOHq) E:\demo > pytest -s --reruns 2 --reruns-delay 3
==================== test session starts ====================
platform win32 -- Python 3.7.9, pytest-7.1.3, pluggy-1.0.0
rootdir: E:\demo
plugins: assume-2.4.3, rerunfailures-10.2
collected 2 items

test_demo.py in test_01
Rin test_01
Rin test_01
Fin test_02
.

========================= FAILURES ==========================
_____ test_01 _____

    def test_01():
        print("in test_01")
>       assert 1 == 2
E       assert 1 == 2

test_demo.py:3: AssertionError
================ short test summary info =================
FAILED test_demo.py::test_01 - assert 1 == 2
========== 1 failed, 1 passed, 2 rerun in 6.07s ==========
```

(demo-HCIhX0Hq) E:\demo>

5.5 如何在一个用例断言失败后继续执行

在自动化测试脚本中，常常有这样的场景，即在一个测试用例中有多个断言，如下所示，在脚本中就存在着三条断言语句。

```
def test_01():
    a = 1
    b = 2
    c = 3
    assert a == 2
    assert b == 3
    assert c == 4
```

直接使用 Pytest 命令执行时，如果执行第一条断言语句报错就会停下来，不再执行后面的断言语句，如下所示。

```
(demo-HCIhX0Hq) E:\demo> pytest
======================== test session starts ========================
platform win32 -- Python 3.7.9, pytest-7.1.3, pluggy-1.0.0
rootdir: E:\demo
plugins: rerunfailures-10.2
collected 1 item

test_demo.py F                                                 [100%]

============================= FAILURES ==============================
_____ test_01 _____

    def test_01():
        a = 1
        b = 2
        c = 3
>       assert a == 2
E       assert 1 == 2

test_demo.py:5: AssertionError
====================== short test summary info ======================
FAILED test_demo.py::test_01 - assert 1 == 2
========================= 1 failed in 0.07s =========================
```

```
(demo-HCIhXOHq) E:\demo>
```

在实际应用中,我们常常希望在一个用例中执行几条断言语句,并对这几条断言内容进行定位分析。要实现这个功能,需要安装一个第三方插件 Pytest-assume,我们执行如下命令安装插件。

```
pip install pytest-assume
```

然后,修改测试脚本,将 assert 更换为 Pytest.assume,如下所示。

```python
import pytest
def test_01():
    a = 1
    b = 2
    c = 3
    pytest.assume(a == 2)
    pytest.assume(b == 3)
    pytest.assume(c == 4)
```

接着,使用 Pytest 命令执行用例,可以看到最终的结果仍然是测试脚本失败,过程中显示三个断言均执行了,且提示有三个断言失败。

```
(demo-HCIhXOHq) E:\demo> pytest
==================== test session starts ====================
platform win32 -- Python 3.7.9, pytest-7.1.3, pluggy-1.0.0
rootdir: E:\demo
plugins: assume-2.4.3, rerunfailures-10.2
collected 1 item

test_demo.py F                                        [100%]


=========================== FAILURES ===========================
_____ test_01 _____

tp = <class 'pytest_assume.plugin.FailedAssumption'>
value = None, tb = None

    def reraise(tp, value, tb=None):
        try:
            if value is None:
                value = tp()
            if value.__traceback__ is not tb:
>               raise value.with_traceback(tb)
E               pytest_assume.plugin.FailedAssumption:
E               3 Failed Assumptions:
```

```
E
E                    test_demo.py:6: AssumptionFailure
E                    >>      pytest.assume(a == 2)
E                    AssertionError: assert False
E
E                    test_demo.py:7: AssumptionFailure
E                    >>      pytest.assume(b == 3)
E                    AssertionError: assert False
E
E                    test_demo.py:8: AssumptionFailure
E                    >>      pytest.assume(c == 4)
E                    AssertionError: assert False

C:\Users\Administrator\.virtualenvs\demo-HCIhX0Hq\lib\site-packages\six.py:
718: FailedAssumption
=================== short test summary info ==================
FAILED test_demo.py::test_01 - pytest_assume.plugin.Faile...
===================== 1 failed in 0.12s =====================

(demo-HCIhX0Hq) E:\demo>
```

如果不太习惯使用 Pytest.assume 进行断言,还可以使用 with pytest.assume 上下文管理器,然后使用 Assert 进行断言即可。

```
from pytest import assume
def test_01():
    a = 1
    b = 2
    c = 3
    with assume: assert a == 2
    with assume: assert b == 3
    with assume: assert c == 4
```

执行结果发现此时仍然是所有的断言均执行。

```
(demo-HCIhX0Hq) E:\demo> pytest
===================== test session starts ====================
platform win32 -- Python 3.7.9, pytest-7.1.3, pluggy-1.0.0
rootdir: E:\demo
plugins: assume-2.4.3, rerunfailures-10.2
collected 1 item

test_demo.py F                                          [100%]
```

```
=========================== FAILURES ===========================
_____ test_01 _____

    def test_01():
        a = 1
        b = 2
        c = 3
        with assume: assert a == 2
        with assume: assert b == 3
>       with assume: assert c == 4
E       pytest_assume.plugin.FailedAssumption:
E       3 Failed Assumptions:
E
E       test_demo.py:6: AssumptionFailure
E       >>      with assume: assert a == 2
E       AssertionError: assert 1 == 2
E
E       test_demo.py:7: AssumptionFailure
E       >>      with assume: assert b == 3
E       AssertionError: assert 2 == 3
E
E       test_demo.py:8: AssumptionFailure
E       >>      with assume: assert c == 4
E       AssertionError: assert 3 == 4

test_demo.py:8: FailedAssumption
================= short test summary info =================
FAILED test_demo.py::test_01 - pytest_assume.plugin.Faile...
==================== 1 failed in 0.10s ====================

(demo-HCIhXOHq) E:\demo>
```

5.6 如何在失败 N 个用例后停止执行

在通常情况下，Pytest 命令是将所有的脚本执行完毕，然后统计通过总数和失败总数。在实践中也发现会有这样的场景，比如有 1 000 个脚本，通常情况下失败的总数不会超过 10 个，但当遇到环境问题或网络问题等时，则会出现大量的脚本失败，消耗了大量的时间。因此，在这种情况下，一般会设置假设 1 000 个脚本执行失败的不会超过 10 个，那么我们可以设置当失败脚本达到 10 个时就停止下来，不再执行，即此时认为可能是环境或网络等出现了问题，有待进一步确认。Pytest 就提供了这样的功能。

如下测试脚本中有 6 个测试用例,其中 test_2、test_4 和 test_6 断言失败。

```python
def test_1():
    print("in test_1")
    assert 1 == 1


def test_2():
    print("in test_2")
    assert 1 == 2


def test_3():
    print("in test_3")
    assert 1 == 1


def test_4():
    print("in test_4")
    assert 1 == 2


def test_5():
    print("in test_5")
    assert 1 == 1


def test_6():
    print("in test_4")
    assert 1 == 2
```

Pytest 可以通过 --maxfail 参数控制最大失败用例的个数,比如设置最大失败用例为 2,即当遇到 2 个失败用例时就停止执行,可以看到,当执行到 test_4 时,便遇到了第二个失败的用例,因此执行完成 test_4 后停下来了,test_5 和 test_6 就不再执行。

```
(demo-HCIhX0Hq) E:\demo> pytest --maxfail 2
==================== test session starts ====================
platform win32 -- Python 3.7.9, pytest-7.1.3, pluggy-1.0.0
rootdir: E:\demo
plugins: assume-2.4.3, rerunfailures-10.2
collected 6 items

test_demo.py .F.F
```

```
========================= FAILURES =========================
_____ test_2 _____

    def test_2():
        print("in test_2")
>       assert 1 == 2
E       assert 1 == 2

test_demo.py:8: AssertionError
------------------- Captured stdout call -------------------
in test_2
_____ test_4 _____

    def test_4():
        print("in test_4")
>       assert 1 == 2
E       assert 1 == 2

test_demo.py:18: AssertionError
------------------- Captured stdout call -------------------
in test_4
================= short test summary info =================
FAILED test_demo.py::test_2 - assert 1 == 2
FAILED test_demo.py::test_4 - assert 1 == 2
!!!!!!!!!!!!!!!! stopping after 2 failures !!!!!!!!!!!!!!!!
================ 2 failed, 2 passed in 0.08s ================

(demo-HCIhXOHq) E:\demo>
```

5.7 如何只执行上次失败的用例

在调试自动化脚本的时候,我们经常遇到的情形是迅速编写了一系列的自动化脚本,然后一起调试,调试的时候,如果直接使用 Pytest 命令,则每次都会把所有的脚本都运行了,而调试时我们希望调通了下次就不运行了,即每次调试执行时只执行上次失败的用例,这样,当没有用例执行时表示所有脚本都已经调试通过了。Pytest 是支持这种执行策略的,只需要在执行的时候增加--lf 参数即可,比如如下三个脚本中,我们设置了 test_01 是通过的,test_02 和 test_03 是断言失败的。

```
def test_1():
    print("in test_1")
    assert 1 == 1
```

```python
def test_2():
    print("in test_2")
    assert 1 == 2


def test_3():
    print("in test_3")
    assert 1 == 3
```

首先,使用 Pytest 命令执行一遍用例,其中,这里 -s 参数是为了显示打印信息。

```
(demo-HCIhXOHq) E:\demo> pytest -s
==================== test session starts ====================
platform win32 -- Python 3.7.9, pytest-7.1.3, pluggy-1.0.0
rootdir: E:\demo
plugins: assume-2.4.3, rerunfailures-10.2
collected 3 items

test_demo.py in test_1
.in test_2
Fin test_3
F


========================== FAILURES ==========================
_____ test_2 _____

    def test_2():
        print("in test_2")
>       assert 1 == 2
E       assert 1 == 2

test_demo.py:8: AssertionError
_____ test_3 _____

    def test_3():
        print("in test_3")
>       assert 1 == 3
E       assert 1 == 3

test_demo.py:13: AssertionError
================== short test summary info ==================
FAILED test_demo.py::test_2 - assert 1 == 2
```

```
FAILED test_demo.py::test_3 - assert 1 == 3
=============== 2 failed, 1 passed in 0.07s ===============

(demo-HCIhXOHq) E:\demo>
```

比如，这里可以发现 test_02 和 test_03 失败了，那么我们把 test_02 中的断言修改正确，如下所示：

```
def test_1():
    print("in test_1")
    assert 1 == 1


def test_2():
    print("in test_2")
    assert 1 == 1


def test_3():
    print("in test_3")
    assert 1 == 3
```

此时，使用 Pytest -s --lf 命令执行用例，同样，-s 参数显示打印的信息，--lf 参数则只执行上次失败的用例。通过打印的信息可以发现此次只执行了 test_02 和 test_03，test_02 通过了，test_03 失败了。

```
(demo-HCIhXOHq) E:\demo> pytest -s --lf
=================== test session starts ===================
platform win32 -- Python 3.7.9, pytest-7.1.3, pluggy-1.0.0
rootdir: E:\demo
plugins: assume-2.4.3, rerunfailures-10.2
collected 2 items
run-last-failure: rerun previous 2 failures

test_demo.py in test_2
. in test_3
F


========================= FAILURES =========================
_____ test_3 _____

    def test_3():
        print("in test_3")
>       assert 1 == 3
```

```
E       assert 1 == 3
```

test_demo.py:13: AssertionError
=================== short test summary info ===================
FAILED test_demo.py::test_3 - assert 1 == 3
================= 1 failed, 1 passed in 0.07s =================

(demo-HCIhXOHq) E:\demo>

我们进一步将 test_03 中的断言也修改正确，如下所示。

```
def test_1():
    print("in test_1")
    assert 1 == 1


def test_2():
    print("in test_2")
    assert 1 == 1


def test_3():
    print("in test_3")
    assert 1 == 1
```

再一次使用 Pytest -s --lf 命令执行，我们看到只执行了上一次失败了的 test_03，这次 test_03 也通过了，至此，所有的脚本均已调试通过了。

```
(demo-HCIhXOHq) E:\demo> pytest -s --lf
==================== test session starts ====================
platform win32 -- Python 3.7.9, pytest-7.1.3, pluggy-1.0.0
rootdir: E:\demo
plugins: assume-2.4.3, rerunfailures-10.2
collected 1 item
run-last-failure: rerun previous 1 failure

test_demo.py in test_3
.

==================== 1 passed in 0.01s ====================

(demo-HCIhXOHq) E:\demo>
```

上述内容就是只执行上一次失败脚本的使用方法以及使用场景。

5.8 如何从上次失败处继续执行用例

Pytest 框架为脚本调试提供了非常多的调试执行策略，本节将继续介绍如何从前次失败地方继续执行用例。

在实际自动化开发调试过程中，很多时候我们希望在调试一批脚本时，一遇到测试用例失败，就停止执行，然后对脚本进行分析修复，从再执行。当再次遇到用例失败时则还要停下来，接着进行定位分析和脚本修复。如此反复进行，最终将所有脚本调试均通过。下面就用脚本演示如何使用这一执行策略。通过分析可以看出，test_01 可以通过，而 test_02 和 test_03 断言失败。

```python
def test_1():
    print("in test_1")
    assert1 == 1


def test_2():
    print("in test_2")
    assert 1 == 2


def test_3():
    print("in test_3")
    assert 1 == 3
```

这里使用 Pytest-s 命令执行一次脚本，结果如下。
test_02 和 test_03 均失败。

```
(demo-HCIhX0Hq) E:\demo> pytest -s
==================== test session starts ====================
platform win32 -- Python 3.7.9, pytest-7.1.3, pluggy-1.0.0
rootdir: E:\demo
plugins: assume-2.4.3, rerunfailures-10.2
collected 3 items

test_demo.py in test_1
.in test_2
Fin test_3
F

============================ FAILURES ============================
```

_____ test_2 _____

```
    def test_2():
        print("in test_2")
>       assert 1 == 2
E       assert 1 == 2
```

test_demo.py:8: AssertionError
_____ test_3 _____

```
    def test_3():
        print("in test_3")
>       assert 1 == 3
E       assert 1 == 3
```

test_demo.py:13: AssertionError
================ short test summary info ================
FAILED test_demo.py::test_2 - assert 1 == 2
FAILED test_demo.py::test_3 - assert 1 == 3
=============== 2 failed, 1 passed in 0.07s ===============

(demo-HCIhXOHq) E:\demo>

对 test_02 脚本进行修复模拟，修复正确，代码如下所示。

```
def test_1():
    print("in test_1")
    assert1 == 1




def test_2():
    print("in test_2")
    assert 1 == 4




def test_3():
    print("in test_3")
    assert 1 == 3
```

使用 Pytest-s--sw 进行调试，其中-s 参数主要用于显示打印信息，--sw 可以从上次失败的地方执行脚本，如果遇到失败则停止。发现 test_02 未修复正确，因此 test_02 仍然失败。

(demo-HCIhXOHq) E:\demo> pytest -s --sw

```
==================== test session starts ====================
platform win32 -- Python 3.7.9, pytest-7.1.3, pluggy-1.0.0
rootdir: E:\demo
plugins: assume-2.4.3, rerunfailures-10.2
collected 3 items
stepwise: no previously failed tests, not skipping.


test_demo.py in test_1
. in test_2
F


======================== FAILURES ========================
_____ test_2 _____

    def test_2():
        print("in test_2")
>       assert 1 == 4
E       assert 1 == 4

test_demo.py:8: AssertionError
================= short test summary info =================
FAILED test_demo.py::test_2 - assert 1 == 4
! Interrupted: Test failed, continuing from this test next run. !
=============== 1 failed, 1 passed in 0.11s ===============


(demo-HCIhXOHq) E:\demo>
```

此时，test_02 脚本修复正确，如下所示。

```
def test_1():
    print("in test_1")
    assert1 == 1


def test_2():
    print("in test_2")
    assert1 == 1


def test_3():
    print("in test_3")
    assert 1 == 3
```

然后继续使用 Pytest -s --sw 执行脚本，即从上一次失败的 test_02 开始执行，我们发现此次 test_02 脚本修复正确了,那么执行 test_03,而 test_03 执行结果又失败了。

```
(demo-HCIhXOHq) E:\demo> pytest -s --sw
==================== test session starts ====================
platform win32 -- Python 3.7.9, pytest-7.1.3, pluggy-1.0.0
rootdir: E:\demo
plugins: assume-2.4.3, rerunfailures-10.2
collected 3 items / 1 deselected / 2 selected
stepwise: skipping 1 already passed items.

test_demo.py in test_2
.in test_3
F

========================= FAILURES =========================
_____ test_3 _____

    def test_3():
        print("in test_3")
>       assert 1 == 3
E       assert 1 == 3

test_demo.py:13: AssertionError
================= short test summary info =================
FAILED test_demo.py::test_3 - assert 1 == 3
! Interrupted: Test failed, continuing from this test next run. !
========1 failed, 1 passed, 1 deselected in 0.11s ========

(demo-HCIhXOHq) E:\demo>
```

继续对 test_03 脚本进行修复,如下所示。

```
def test_1():
    print("in test_1")
    assert1 == 1



def test_2():
    print("in test_2")
    assert1 == 1



def test_3():
    print("in test_3")
```

```
assert1 == 1
```

然后使用 Pytest -s --sw 命令继续执行,此时,则 test_03 也修复正确了,执行结果如下。

```
(demo-HCIhX0Hq) E:\demo> pytest -s --sw
==================== test session starts ====================
platform win32 -- Python 3.7.9, pytest-7.1.3, pluggy-1.0.0
rootdir: E:\demo
plugins: assume-2.4.3, rerunfailures-10.2
collected 3 items / 2 deselected / 1 selected
stepwise: skipping 2 already passed items.

test_demo.py in test_3
.

============= 1 passed, 2 deselected in 0.02s =============

(demo-HCIhX0Hq) E:\demo>
```

上述内容表述了如何从上次执行失败的地方继续执行脚本的使用方法和使用场景。

5.9 如何先执行上次失败用例,再执行其他用例

在对自动化脚本连跑测试过程中,往往希望优先执行上次连跑失败的脚本,然后再执行其他脚本,Pytest 框架就支持这样一种连跑机制,通过 --ff 参数即可完成。我们准备三个脚本,即第一个和第三个断言失败,第二个断言成功。

```
def test_01():
    assert 1 == 2

def test_02():
    assert 2 == 2

def test_03():
    assert 1 == 2
```

先执行一次脚本,我们可以看出,此时的执行顺序是按照 test_01、test_02、test_03 的顺序执行的。

```
(demo-- ip5ZZo0) D:\demo> pytest
==================== test session starts ====================
```

```
platform win32 -- Python 3.7.9, pytest-7.1.3, pluggy-1.0.0
rootdir: D:\demo
collected 3 items

test_demo.py F.F                                              [100%]


========================= FAILURES =========================
_____ test_01 _____

    def test_01():
>       assert 1 == 2
E       assert 1 == 2

test_demo.py:3: AssertionError
_____ test_03 _____

    def test_03():
>       assert 1 == 2
E       assert 1 == 2

test_demo.py:9: AssertionError
================= short test summary info =================
FAILED test_demo.py::test_01 - assert 1 == 2
FAILED test_demo.py::test_03 - assert 1 == 2
================ 2 failed, 1 passed in 0.18s ================

(demo--ip5ZZo0) D:\demo>
```

下面使用 Pytest --ff 参数再次执行用例,可以看出,此时先执行 test_01 和 test_03,最后执行了 test_02,即首先执行了上次失败的用例,然后才执行其他用例。

```
(demo--ip5ZZo0) D:\demo> pytest --ff
==================== test session starts ====================
platform win32 -- Python 3.7.9, pytest-7.1.3, pluggy-1.0.0
rootdir: D:\demo
collected 3 items
run-last-failure: rerun previous 2 failures first

test_demo.py FF.                                              [100%]


========================= FAILURES =========================
_____ test_01 _____
```

```
    def test_01():
>       assert 1 == 2
E       assert 1 == 2

test_demo.py:3: AssertionError
_____ test_03 _____

    def test_03():
>       assert 1 == 2
E       assert 1 == 2

test_demo.py:9: AssertionError
================ short test summary info ================
FAILED test_demo.py::test_01 - assert 1 == 2
FAILED test_demo.py::test_03 - assert 1 == 2
=============== 2 failed, 1 passed in 0.05s ===============

(demo--ip5ZZo0) D:\demo>
```

5.10 如何重复执行用例

在有些场景下,我们希望将自动化脚本重复执行多次,以察看自动化脚本的稳定性,或者想利用功能测试用例直接进行压力测试,那么在 Pytest 框架中可以通过第三方插件 pytest-repeat 来实现这样的需求。

首先,执行如下命令安装 pytest-repeat 插件。

```
pip install pytest-repeat
```

在脚本中直接使用 @pytest.mark.repeat(N) 标记编写脚本,即在 test_01 上使用 @pytest.mark.repeat(3) 标记。

```
import pytest

@pytest.mark.repeat(3)
def test_01():
    print("in test01...")
    assert 1 == 1

def test_02():
    print("in test02...")
    assert 1 == 1
```

通过打印可以看出，这里 test_01 执行了 3 次。

```
(demo--ip5ZZo0) D:\demo> pytest
==================== test session starts ====================
platform win32 -- Python 3.7.9, pytest-7.1.3, pluggy-1.0.0
rootdir: D:\demo
plugins: repeat-0.9.1
collected 4 items

test_demo.py ....                                    [100%]

==================== 4 passed in 0.03s ====================
(demo--ip5ZZo0) D:\demo>
```

在实际应用中，这种用法不是很多，因为在通常情况下，脚本是不需要执行多次的，所以在执行脚本的命令行中指定重复执行用例往往用得最多。比如，我们在这里将测试脚本修改如下：

```python
def test_01():
    print("in test01...")
    assert 1 == 1

def test_02():
    print("in test02...")
    assert 1 == 1
```

在命令行中使用 -count=N 来重复执行脚本，此时，将 test_01 和 test_02 各执行了 3 次。

```
(demo--ip5ZZo0) D:\demo> pytest -s --count=3
==================== test session starts ====================
platform win32 -- Python 3.7.9, pytest-7.1.3, pluggy-1.0.0
rootdir: D:\demo
plugins: repeat-0.9.1
collected 6 items

test_demo.py in test01...
.in test01...
.in test01...
.in test02...
.in test02...
.in test02...
.
```

==================== 6 passed in 0.03s ====================

(demo -- ip5ZZo0) D:\demo>

从执行结果可以看出,过程中将 test_01 连续执行了 3 次,然后又将 test_02 执行了 3 次,那么这就有一个问题,能不能将 test_01 和 test_02 先均执行完一次,再反复执行,答案是肯定的,这就需要另外一个参数--repeat-scope 出场了。--repeat-scope 支持可选的参数有 function、class、module、session,在前面没有指定-repeat-scope 的情况下-repeat-scope 同 function 执行的是一致的。

(demo -- ip5ZZo0) D:\demo> pytest -s --count=3 --repeat-scope=function
==================== test session starts ====================
platform win32 -- Python 3.7.9, pytest-7.1.3, pluggy-1.0.0
rootdir: D:\demo
plugins: repeat-0.9.1
collected 6 items

test_demo.py in test01...
.in test01...
.in test01...
.in test02...
.in test02...
.in test02...
.

==================== 6 passed in 0.03s ====================

(demo -- ip5ZZo0) D:\demo>

通过指定-repeat-scope=module 即可将 test_01 和 test_02 作为一个整体循环执行多次,执行结果如下所示。

(demo -- ip5ZZo0) D:\demo> pytest -s --count=3 --repeat-scope=module
==================== test session starts ====================
platform win32 -- Python 3.7.9, pytest-7.1.3, pluggy-1.0.0
rootdir: D:\demo
plugins: repeat-0.9.1
collected 6 items

test_demo.py in test01...
.in test02...
.in test01...
.in test02...
.in test01...

.in test02...

==================== 6 passed in 0.02s ====================

(demo -- ip5ZZo0) D:\demo>

将测试脚本使用测试类组织的情况,如下所示。

```
class TestDemo1():
    def test_01(self):
        print("in test01...")
        assert 1 == 1

    def test_02(self):
        print("in test02...")
        assert 1 == 1


class TestDemo2():
    def test_03(self):
        print("in test03...")
        assert 1 == 1

    def test_04(self):
        print("in test04...")
        assert 1 == 1
```

接下来,通过--repeat-scope=class 来指定测试类作为重复执行的整体,执行结果如下所示。

```
(demo -- ip5ZZo0) D:\demo> pytest -s --count=3 --repeat-scope=class
==================== test session starts ====================
platform win32 -- Python 3.7.9, pytest-7.1.3, pluggy-1.0.0
rootdir: D:\demo
plugins: repeat-0.9.1
collected 12 items

test_demo.py in test01...
.in test02...
.in test01...
.in test02...
.in test01...
.in test02...
.in test03...
```

```
.in test04...
.in test03...
.in test04...
.in test03...
.in test04...
.

==================== 12 passed in 0.16s ====================
```

(demo -- ip5ZZo0) D:\demo>

当-repeat-scope=session 时,编写 test_demo.py 和 test_demo2.py 两个测试文件,其中 test_demo.py 文件中的测试代码如下所示。

```
class TestDemo1():
    def test_01(self):
        print("in test01...")
        assert 1 == 1

    def test_02(self):
        print("in test02...")
        assert 1 == 1
```

接着,我们再看 test_demo2.py 文件中的测试代码。

```
class TestDemo2():
    def test_03(self):
        print("in test03...")
        assert 1 == 1

    def test_04(self):
        print("in test04...")
        assert 1 == 1
```

通过指定-repeat-scope=session 将所有脚本作为一个整体重复执行了 3 次。

```
(demo -- ip5ZZo0) D:\demo> pytest -s --count=3 --repeat-scope=session
==================== test session starts ====================
platform win32 -- Python 3.7.9, pytest-7.1.3, pluggy-1.0.0
rootdir: D:\demo
plugins: repeat-0.9.1
collected 12 items

test_demo.py in test01...
.in test02...
```

```
           .
           test_demo2.py in test03...
           .in test04...
           .
           test_demo.py in test01...
           .in test02...
           .
           test_demo2.py in test03...
           .in test04...
           .
           test_demo.py in test01...
           .in test02...
           .
           test_demo2.py in test03...
           .in test04...
           .

           ==================== 12 passed in 2.38s ====================

(demo -- ip5ZZo0) D:\demo >
```

5.11 如何进行多进程并行执行用例

当自动化脚本数量非常多的时候，全量脚本执行耗时很长，此时，我们就希望自动化脚本能够并发执行，第三方插件 Pytest-xdist 能很好支持此功能，Pytest-xdist 是从进程层面真正做到并行执行。

首先安装 Pytest-xdist 插件，执行命令。

```
pip install pytest-xdist
```

下面编写如下脚本，有 8 个测试用例，每个用例中通过 time.sleep 等待模拟用例耗时，具体代码如下，通过粗略估计，如果直接使用 Pytest 顺序执行，至少需要 36 s 的时间。

```python
import time

def test_01():
    time.sleep(1)
    print("in test01...")
    assert 1 == 1


def test_02():
```

```python
    time.sleep(2)
    print("in test02...")
    assert 1 == 1


def test_03():
    time.sleep(3)
    print("in test03...")
    assert 1 == 1


def test_04():
    time.sleep(4)
    print("in test04...")
    assert 1 == 1


def test_05():
    time.sleep(5)
    print("in test05...")
    assert 1 == 1


def test_06():
    time.sleep(6)
    print("in test06...")
    assert 1 == 1


def test_07():
    time.sleep(7)
    print("in test07...")
    assert 1 == 1


def test_08():
    time.sleep(8)
    print("in test08...")
    assert 1 == 1
```

使用 Pytest 命令顺序执行,我们发现总耗时 38 s。

```
(demo-- ip5ZZo0) D:\demo> pytest -s
==================== test session starts ====================
platform win32 -- Python 3.7.9, pytest-7.1.3, pluggy-1.0.0
rootdir: D:\demo
plugins: repeat-0.9.1
collected 8 items

test_demo.py in test01...
```

```
.in test02...
.in test03...
.in test04...
.in test05...
.in test06...
.in test07...
.in test08...
.

==================== 8 passed in 38.36s ====================

(demo -- ip5ZZo0) D:\demo>
```

在安装 Pytest-xdist 插件后,通过 -n auto 参数即可实现并行执行,-n 后面可以指定并行执行的数量,比如 -n 3 表示同一时间有三个任务在执行,-n auto 则会自动计算当前服务器的 cpu 的总核数,将所有的 cpu 全部用上,此时服务器将会出现比较卡的状况。下面用 -n auto 的执行仅需要 9 t,其中在 test_08 中等待了 8 t,是耗时依赖最大耗时的用例。

```
(demo -- ip5ZZo0) D:\demo> pytest -s -n auto
==================== test session starts ====================
platform win32 -- Python 3.7.9, pytest-7.1.3, pluggy-1.0.0
rootdir: D:\demo
plugins: forked-1.4.0, repeat-0.9.1, xdist-2.5.0
gw0 I / gw1 I / gw2 I / gw3 I / gw4 I / gw5 I / gw6 I / gw7 gw0 C / gw1 I / gw2 I / gw3 I / gw4
I / gw5 I / gw6 I / gw7 gw0 C / gw1 C / gw2 C / gw3 I / gw4 C / gw5 I / gw6 I / gw7 gw0 C / gw1 / gw2
C / gw3 I / gw4 I / gw5 I / gw6 I / gw7 gw0 C / gw1 C / gw2 C / gw3 I / gw4 I / gw5 I / gw6 I / gw7 gw0
C / gw1 C / gw2 C / gw3 C / gw4 C / gw5 I / gw6 I / gw7 gw0 C / gw1 C / gw2 C / gw3 C / gw4 C / gw5 C
/ gw6 I / gw7 gw0 C / gw1 C / gw2 C / gw3 C / gw4 C / gw5 C / gw6 C / gw7 gw0 C / gw1 C / gw2 C / gw3
C / gw4 C / gw5 C / gw6 C / gw7 gw0 ok / gw1 C / gw2 C / gw3 C / gw4 C / gw5 C / gw6 C / gw7gw0 [8] /
gw1 C / gw2 C / gw3 C / gw4 C / gw5 C / gw6 C / gwgw0 [8] / gw1 ok / gw2 C / gw3 C / gw4 C / gw5 C / gw6
C / ggw0 [8] / gw1 [8] / gw2 C / gw3 C / gw4 C / gw5 C / gw6 C / gw0 [8] / gw1 [8] / gw2 ok / gw3 C /
gw4 C / gw5 C / gw6 C /gw0 [8] / gw1 [8] / gw2 [8] / gw3 C / gw4 C / gw5 C / gw6 C gw0 [8] / gw1 [8]
/ gw2 [8] / gw3 ok / gw4 C / gw5 C / gw6 Cgw0 [8] / gw1 [8] / gw2 [8] / gw3 [8] / gw4 C / gw5 C / gw6
gw0 [8] / gw1 [8] / gw2 [8] / gw3 [8] / gw4 ok / gw5 C / gw6gw0 [8] / gw1 [8] / gw2 [8] / gw3 [8] /
gw4 [8] / gw5 C / gwgw0 [8] / gw1 [8] / gw2 [8] / gw3 [8] / gw4 [8] / gw5 ok / ggw0 [8] / gw1 [8] /
gw2 [8] / gw3 [8] / gw4 [8] / gw5 [8] / gw0 [8] / gw1 [8] / gw2 [8] / gw3 [8] / gw4 [8] / gw5 [8]
/ gw0 [8] / gw1 [8] / gw2 [8] / gw3 [8] / gw4 [8] / gw5 [8] / gw6 [8] / gw7 [8]
/ gw4 [8] / gw5 [8] / gw0 [8] / gw1 [8] / gw2 [8] / gw3 [8] / gw4 [8] / gw5 [8] / gw6 [8] / gw7 [8]
........
==================== 8 passed in 9.43s ====================

(demo -- ip5ZZo0) D:\demo>
```

我们设置同时执行 4 个任务,即并行执行量设置为 4,此时耗时 12.87 s。

```
(demo -- ip5ZZo0) D:\demo> pytest -s -n 4
==================== test session starts ====================
platform win32 -- Python 3.7.9, pytest-7.1.3, pluggy-1.0.0
rootdir: D:\demo
plugins: forked-1.4.0, repeat-0.9.1, xdist-2.5.0
gw0 [8] / gw1 [8] / gw2 [8] / gw3 [8]
........
==================== 8 passed in 12.87s ====================

(demo -- ip5ZZo0) D:\demo>
```

从上述内容中我们发现通过 Pytest-xdist 插件可以大幅提升脚本的执行效率。

5.12 如何随机执行用例

在编写自动化脚本时,一般都要求用例之间是独立的,期望在最终连跑验证的时候能随机执行,因为随机执行更能模拟手工测试的场景。Pytest 框架通过提供第三方插件 Pytest-random-order 可满足随机执行用例的要求。

执行如下命令,安装第三方插件。

```
pip install pytest-random-order
```

编写如下三个用例的测试脚本。

```
def test_01():
    print("in test01...")
    assert 1 == 1


def test_02():
    print("in test02...")
    assert 1 == 1


def test_03():
    print("in test03...")
    assert 1 == 1
```

使用 -radom-order 参数即可实现自动化脚本的随机执行,通过打印结果可以发现先执行 test_03,然后执行 test_02,最后执行 test_01。

```
(demo -- ip5ZZo0) D:\demo> pytest -s --random-order
==================== test session starts ====================
```

```
platform win32 -- Python 3.7.9, pytest-7.1.3, pluggy-1.0.0
Using --random-order-bucket=module
Using --random-order-seed=937307

rootdir: D:\demo
plugins: forked-1.4.0, random-order-1.0.4, repeat-0.9.1, xdist-2.5.0
collected 3 items

test_demo.py in test03...
.in test02...
.in test01...
.

==================== 3 passed in 2.26s ====================

(demo--ip5ZZo0) D:\demo>
```

此外，Pytest--random-order 插件还可以设置一定范围内随机执行模式，可通过--random-order-bucket 的参数设置范围，可设置的范围有 module 和 class，其他选项基本用不着，因此这里只介绍 module 和 class 两个选项。为了更好演示随机性，我们重新编写自动化脚本，这里一个测试文件中还有类外的测试函数，有两个测试类，每个测试类中有三个测试方法。

```python
def test_01():
    print("in test01...")
    assert 1 == 1

def test_02():
    print("in test02...")
    assert 1 == 1

def test_03():
    print("in test03...")
    assert 1 == 1

class TestDemo1():
    def test_04(self):
        print("in test04...")
        assert 1 == 1

    def test_05(self):
        print("in test05...")
        assert 1 == 1
```

```python
    def test_06(self):
        print("in test06...")
        assert 1 == 1

class TestDemo2():
    def test_07(self):
        print("in test07...")
        assert 1 == 1

    def test_08(self):
        print("in test08...")
        assert 1 == 1

    def test_09(self):
        print("in test09...")
        assert 1 == 1
```

我们使用--random-order参数不指定随机范围的时候执行用例,可以发现此时所有用例都是随机执行的。

```
(demo--ip5ZZo0) D:\demo> pytest -s --random-order
==================== test session starts ====================
platform win32 -- Python 3.7.9, pytest-7.1.3, pluggy-1.0.0
Using --random-order-bucket=module
Using --random-order-seed=982647

rootdir: D:\demo
plugins: forked-1.4.0, random-order-1.0.4, repeat-0.9.1, xdist-2.5.0
collected 9 items

test_demo.py in test05...
.in test04...
.in test02...
.in test03...
.in test07...
.in test06...
.in test09...
.in test01...
.in test08...
.

==================== 9 passed in 0.03s ====================
```

```
(demo--ip5ZZo0) D:\demo>
```

--random-order-bucket=class 参数可以指定每个类中的测试方法随机执行,此时一个类中的方法是随机的,这种执行方法对测试类中有 setup 和 teardown 的用例是非常有用的。

```
(demo--ip5ZZo0) D:\demo> pytest -s --random-order-bucket=class
==================== test session starts ====================
platform win32 -- Python 3.7.9, pytest-7.1.3, pluggy-1.0.0
Using --random-order-bucket=class
Using --random-order-seed=186060

rootdir: D:\demo
plugins: forked-1.4.0, random-order-1.0.4, repeat-0.9.1, xdist-2.5.0
collected 9 items

test_demo.py in test01...
.in test02...
.in test03...
.in test07...
.in test08...
.in test09...
.in test05...
.in test04...
.in test06...
.
==================== 9 passed in 0.04s ====================

(demo--ip5ZZo0) D:\demo>
```

我们看,如果设置模块内随机,由于只有一个测试文件,因此测试文件内所有用例都随机执行了。而当存在多个测试文件时,每个测试文件内也是随机执行的。

```
(demo--ip5ZZo0) D:\demo> pytest -s --random-order-bucket=module
==================== test session starts ====================
platform win32 -- Python 3.7.9, pytest-7.1.3, pluggy-1.0.0
Using --random-order-bucket=module
Using --random-order-seed=91396

rootdir: D:\demo
plugins: forked-1.4.0, random-order-1.0.4, repeat-0.9.1, xdist-2.5.0
collected 9 items
```

```
test_demo.py in test01...
.in test08...
.in test09...
.in test05...
.in test04...
.in test07...
.in test03...
.in test02...
.in test06...
.

=================== 9 passed in 0.04s ===================

(demo -- ip5ZZo0) D:\demo>
```

其实,在随机执行脚本的时候也会有新的问题,比如当第一次随机执行脚本时,如果中间有脚本失败了,再次重新随机执行时就不会失败了,因为某些用例在特定的执行顺序下才会出现,但这种定位有些困难。Pytest--random-order插件可通过使用-random--order-seed=N 的参数使执行的顺序和前一次保持一致,下面是第一执行指定 N 为 10000 的执行结果。

```
(demo -- ip5ZZo0) D:\demo> pytest -s --random-order-seed=10000
=================== test session starts ===================
platform win32 -- Python 3.7.9, pytest-7.1.3, pluggy-1.0.0
Using --random-order-bucket=module
Using --random-order-seed=10000

rootdir: D:\demo
plugins: forked-1.4.0, picked-0.4.6, random-order-1.0.4, repeat-0.9.1, xdist-2.5.0
collected 9 items

test_demo.py in test06...
.in test08...
.in test04...
.in test05...
.in test07...
.in test03...
.in test09...
.in test02...
.in test01...
.

=================== 9 passed in 0.04s ===================
```

```
(demo -- ip5ZZo0) D:\demo>
```

我们看到,当再次使用 Pytest -s --random-order-seed=10 000 命令执行,此时的执行顺序和上一次执行完全一致。

```
(demo -- ip5ZZo0) D:\demo> pytest -s --random-order-seed=10000
==================== test session starts ====================
platform win32 -- Python 3.7.9, pytest-7.1.3, pluggy-1.0.0
Using --random-order-bucket=module
Using --random-order-seed=10000

rootdir: D:\demo
plugins: forked-1.4.0, picked-0.4.6, random-order-1.0.4, repeat-0.9.1, xdist-2.5.0
collected 9 items

test_demo.py in test06...
.in test08...
.in test04...
.in test05...
.in test07...
.in test03...
.in test09...
.in test02...
.in test01...
.
==================== 9 passed in 0.04s ====================

(demo -- ip5ZZo0) D:\demo>
```

5.13 如何只运行未提交 git 代码仓的脚本

当使用 git 管理脚本代码时,如果我们想只跑尚未提交的代码仓和在本地已经修改的脚本,以及优先执行新增或修改的脚本,就需要使用 Pytest 的第三方插件 Pytest-picked 了来完成。我们这里先使用如下命令安装一下插件。

```
pip install pytest-picked
```

接下来,准备一下测试环境,在 test_demo.py 文件中写入如下所示测试代码。

```
def test_01():
    print("in test01...")
```

```
    assert 1 == 1

def test_02():
    print("in test02...")
    assert 1 == 1
```

然后,执行 git init 命令,将当前目录初始化为 git 代码仓,执行结果如下所示。

```
(demo -- ip5ZZo0) D:\demo> git init
warning: templates not found in .git_template
Initialized empty Git repository in D:/demo/.git/

(demo -- ip5ZZo0) D:\demo>
```

之后使用 git status 查看当前仓库的状态,即此时为 test_demo.py。

```
(demo -- ip5ZZo0) D:\demo> git status
On branch master
Untracked files:
    (use "git add <file>..." to include in what will be committed)
        test_demo.py

nothing added to commit but untracked files present (use "git add" to track)

(demo -- ip5ZZo0) D:\demo>
```

将 test_demo.py 提交到仓库,我们看到,通过 git add 和 git commit 之后已经全部提交到仓库了。

```
(demo -- ip5ZZo0) D:\demo> git add .

(demo -- ip5ZZo0) D:\demo> git commit -m "add test_demo.py"
[master beff6ce] add test_demo.py
 1 file changed, 10 insertions(+)
 create mode 100644 test_demo.py

(demo -- ip5ZZo0) D:\demo> git status
On branch master
nothing to commit, working tree clean

(demo -- ip5ZZo0) D:\demo>
```

Pytest 通过使用 -picked 参数即可执行未提交到仓库的代码,此时先使用 Pytest-picked 命令执行一次用例,结果提示没有测试脚本。

```
(demo -- ip5ZZo0) D:\demo> pytest -s --picked
```

```
Changed test files... 0.
Changed test folders... 0.
=================== test session starts ===================
platform win32 -- Python 3.7.9, pytest-7.1.3, pluggy-1.0.0
Test order randomisation NOT enabled. Enable with --random-order or --random-order-bucket=<bucket_type>
rootdir: D:\demo, testpaths:
plugins: forked-1.4.0, picked-0.4.6, random-order-1.0.4, repeat-0.9.1, xdist-2.5.0
collected 0 items

================== no tests ran in 0.01s ==================

(demo -- ip5ZZo0) D:\demo>
```

这里，新增了一个测试文件 test_demo2.py，编写测试代码如下所示。

```
def test_03():
    print("in test03...")
    assert 1 == 1

def test_04():
    print("in test04...")
    assert 1 == 1
```

通过 git status 命令我们可以看到此时 test_demo2.py 处于尚未提交的状态，如下所示。

```
(demo -- ip5ZZo0) D:\demo> git status
On branch master
Untracked files:
    (use "git add <file>..." to include in what will be committed)
        test_demo2.py

nothing added to commit but untracked files present (use "git add" to track)

(demo -- ip5ZZo0) D:\demo>
```

再次执行 Pytest -s -picked 命令，可以发现，此时只执行了尚未提交的测试用例 test_03 和 test_04，而已经提交的 test_demo.py 文件中的 test_01 和 test_02 未被执行。说明达到了预期的效果。

当然在有些时候，我们也有可能希望优先执行尚未提交的测试脚本，然后再执行其他的脚本，Pytest-picked 插件也是支持此功能的，实际操作时只需要指定 -picked=first 即可满足。通过打印的结果可以看出，此时先执行了 test_03 和 test_04，然后又执行 test_01 和 test_02。

```
(demo--ip5ZZo0) D:\demo> pytest -s --picked=first
==================== test session starts ====================
platform win32 -- Python 3.7.9, pytest-7.1.3, pluggy-1.0.0
Test order randomisation NOT enabled. Enable with --random-order or --random-order-bucket=<bucket_type>
rootdir: D:\demo
plugins: forked-1.4.0, picked-0.4.6, random-order-1.0.4, repeat-0.9.1, xdist-2.5.0
collected 4 items

test_demo2.py in test03...
.in test04...
.
test_demo.py in test01...
.in test02...
.

==================== 4 passed in 2.48s ====================

(demo--ip5ZZo0) D:\demo>
```

上述就是对 Pytest-picked 插件的主要功能的介绍，即可以只执行尚未提交 git 代码仓库的测试脚本或者优先执行尚未提交 git 代码仓的脚本。

5.14 如何查找耗时最长的用例脚本

当存在大量测试脚本的时候，脚本的运行耗时就会变得很长，难免需要对脚本的运行时间进行分析，而要想提高自动化脚本的运行效率，最直接、最有效的方法就是找出耗时最大的几个脚本，然后对其进行性能优化，那么怎么找出耗时大的 N 个脚本呢。

Pytest 框架提供了这样的功能，可通过 -durations 参数对脚本的运行时间进行分析，编写多个测试脚本，然后使用 sleep 模拟脚本的耗时。

```
import time

def test_01():
    print("in test01...")
    time.sleep(1)
    assert 1 == 1

def test_02():
```

```python
        print("in test02...")
        time.sleep(2)
        assert 1 == 1

    def test_03():
        print("in test03...")
        time.sleep(3)
        assert 1 == 1

    def test_04():
        print("in test04...")
        time.sleep(4)
        assert 1 == 1

    def test_05():
        print("in test05...")
        time.sleep(5)
        assert 1 == 1

    def test_06():
        print("in test06...")
        time.sleep(6)
        assert 1 == 1

    def test_07():
        print("in test07...")
        time.sleep(7)
        assert 1 == 1

    def test_08():
        print("in test08...")
        time.sleep(8)
        assert 1 == 1

    def test_09():
        print("in test09...")
        time.sleep(9)
        assert 1 == 1

    def test_10():
        print("in test10...")
        time.sleep(10)
        assert 1 == 1
```

当使用 pytest-durations=0,可将所有脚本的耗时打印出来,执行结果的 slowest durations 的部分会显示所有用例的执行耗时。

```
(demo -- ip5ZZo0) D:\demo> pytest -s --durations=0
=================== test session starts ===================
platform win32 -- Python 3.7.9, pytest-7.1.3, pluggy-1.0.0
Test order randomisation NOT enabled. Enable with --random-order or --random-order-bucket=<bucket_type>
rootdir: D:\demo
plugins: forked-1.4.0, picked-0.4.6, random-order-1.0.4, repeat-0.9.1, xdist-2.5.0
collected 10 items

test_demo.py in test01...
.in test02...
.in test03...
.in test04...
.in test05...
.in test06...
.in test07...
.in test08...
.in test09...
.in test10...
.

=================== slowest durations ===================
10.01s call     test_demo.py::test_10
9.02s call      test_demo.py::test_09
8.01s call      test_demo.py::test_08
7.01s call      test_demo.py::test_07
6.02s call      test_demo.py::test_06
5.01s call      test_demo.py::test_05
4.01s call      test_demo.py::test_04
3.00s call      test_demo.py::test_03
2.00s call      test_demo.py::test_02
1.00s call      test_demo.py::test_01

(20 durations < 0.005s hidden.  Use -vv to show these durations.)
=================== 10 passed in 57.40s ===================

(demo -- ip5ZZo0) D:\demo>
```

如用例很多,我们只想找出耗时最大的 5 个脚本,那么使用 Pytest-durations=5 即可找到。我们可以看出,此时只显示了耗时最大的 5 个测试脚本的耗时,在实际开发应用中,可直接分析这 5 个脚本的耗时原因,然后对其进行优化即可。

```
(demo--ip5ZZo0) D:\demo> pytest --durations=5
==================== test session starts ====================
platform win32 -- Python 3.7.9, pytest-7.1.3, pluggy-1.0.0
Test order randomisation NOT enabled. Enable with --random-order or --random-order-bucket=<bucket_type>
rootdir: D:\demo
plugins: forked-1.4.0, picked-0.4.6, random-order-1.0.4, repeat-0.9.1, xdist-2.5.0
collected 10 items

test_demo.py ..........                                   [100%]

==================== slowest 5 durations ====================
10.01s call     test_demo.py::test_10
9.02s call      test_demo.py::test_09
8.00s call      test_demo.py::test_08
7.01s call      test_demo.py::test_07
6.01s call      test_demo.py::test_06
==================== 10 passed in 55.14s ====================

(demo--ip5ZZo0) D:\demo>
```

这里要注意,就是当脚本耗时小于 0.005 s 时,使用--durations=0 也不会显示用例执行。

```
def test_01():
    print("in test01...")
    assert 1 == 1

def test_02():
    print("in test02...")
    assert 1 == 1

def test_03():
    print("in test03...")
    assert 1 == 1

def test_04():
    print("in test04...")
    assert 1 == 1

def test_05():
    print("in test05...")
```

```
        assert 1 == 1

    def test_06():
        print("in test06...")
        assert 1 == 1

    def test_07():
        print("in test07...")
        assert 1 == 1

    def test_08():
        print("in test08...")
        assert 1 == 1

    def test_09():
        print("in test09...")
        assert 1 == 1

    def test_10():
        print("in test10...")
        assert 1 == 1
```

这里使用--durations执行完成后,结果也不显示用例的执行耗时。

```
(demo -- ip5ZZo0) D:\demo> pytest -- durations = 0
=================== test session starts ===================
platform win32 -- Python 3.7.9, pytest-7.1.3, pluggy-1.0.0
Test order randomisation NOT enabled. Enable with --random-order or --random-order-bucket=<bucket_type>
rootdir: D:\demo
plugins: forked-1.4.0, picked-0.4.6, random-order-1.0.4, repeat-0.9.1, xdist-2.5.0
collected 10 items

test_demo.py ..........                                    [100%]

=================== slowest durations ===================

(30 durations < 0.005s hidden.  Use -vv to show these durations.)
=================== 10 passed in 2.19s ===================

(demo -- ip5ZZo0) D:\demo>
```

我们看到,在执行的最后提示用例耗时小于0.005 s,如果在这种情况下还要显示用例的耗时时间,加上-vv参数即可,执行结果为:这里虽然显示了耗时,但是都显示了

0.00 s。因此在一般情况下,当耗时小于 0.005 s 时,基本可以忽略不计了。

```
(demo--ip5ZZo0) D:\demo> pytest --durations=0 -vv
==================== test session starts ====================
platform win32 -- Python 3.7.9, pytest-7.1.3, pluggy-1.0.0 -- C:\Users\hitre\.virtualenvs\demo--ip5ZZo0\Scripts\python.exe
cachedir: .pytest_cache
Test order randomisation NOT enabled. Enable with --random-order or --random-order-bucket=<bucket_type>
rootdir: D:\demo
plugins: forked-1.4.0, picked-0.4.6, random-order-1.0.4, repeat-0.9.1, xdist-2.5.0
collected 10 items

test_demo.py::test_01 PASSED                                    [ 10%]
test_demo.py::test_02 PASSED                                    [ 20%]
test_demo.py::test_03 PASSED                                    [ 30%]
test_demo.py::test_04 PASSED                                    [ 40%]
test_demo.py::test_05 PASSED                                    [ 50%]
test_demo.py::test_06 PASSED                                    [ 60%]
test_demo.py::test_07 PASSED                                    [ 70%]
test_demo.py::test_08 PASSED                                    [ 80%]
test_demo.py::test_09 PASSED                                    [ 90%]
test_demo.py::test_10 PASSED                                    [100%]

==================== slowest durations ====================
0.00s call       test_demo.py::test_01
0.00s call       test_demo.py::test_04
0.00s call       test_demo.py::test_05
0.00s call       test_demo.py::test_02
0.00s call       test_demo.py::test_07
0.00s call       test_demo.py::test_06
0.00s call       test_demo.py::test_10
0.00s call       test_demo.py::test_08
0.00s call       test_demo.py::test_03
0.00s call       test_demo.py::test_09
0.00s teardown   test_demo.py::test_01
0.00s setup      test_demo.py::test_01
0.00s teardown   test_demo.py::test_05
0.00s setup      test_demo.py::test_04
0.00s setup      test_demo.py::test_02
0.00s setup      test_demo.py::test_05
0.00s setup      test_demo.py::test_07
0.00s teardown   test_demo.py::test_04
```

```
0.00s teardown     test_demo.py::test_08
0.00s teardown     test_demo.py::test_02
0.00s teardown     test_demo.py::test_03
0.00s setup        test_demo.py::test_06
0.00s setup        test_demo.py::test_09
0.00s teardown     test_demo.py::test_09
0.00s setup        test_demo.py::test_08
0.00s teardown     test_demo.py::test_10
0.00s setup        test_demo.py::test_03
0.00s teardown     test_demo.py::test_07
0.00s teardown     test_demo.py::test_06
0.00s setup        test_demo.py::test_10
=================== 10 passed in 0.05s ===================

(demo--ip5ZZo0) D:\demo>
```

此外，我们还可以同时指定--durations＝0 和--durations-min＝6 来显示运行时间超过 6 s 的脚本。

```
(demo--ip5ZZo0) D:\demo> pytest --durations=0 --durations-min=6
=================== test session starts ===================
platform win32 -- Python 3.7.9, pytest-7.1.3, pluggy-1.0.0
Test order randomisation NOT enabled. Enable with --random-order or --random-order-bucket=<bucket_type>
rootdir: D:\demo
plugins: forked-1.4.0, picked-0.4.6, random-order-1.0.4, repeat-0.9.1, xdist-2.5.0
collected 10 items

test_demo.py ..........                                          [100%]

=================== slowest durations ===================
10.01s call     test_demo.py::test_10
9.02s call      test_demo.py::test_09
8.00s call      test_demo.py::test_08
7.01s call      test_demo.py::test_07
6.00s call      test_demo.py::test_06

(25 durations <6s hidden.  Use -vv to show these durations.)
=================== 10 passed in 55.15s ===================

(demo--ip5ZZo0) D:\demo>
```

第 6 章　fixture 的基础应用

在 Pytest 中,fixture 一直被认为是比较晦涩难懂的功能,本章就从 fixture 最简单的传值功能开始介绍,一步一步揭开 fixture 的神秘面纱。

6.1　fixture 传值的作用

大家知道,fixture 是 Pytest 中的一个非常重要的功能,同时又被普遍认为是一个非常难懂的功能,我们就从 fixture 最简单的功能讲起。

为了更好讲解 fixture 的功能,我们直接以代码为例介绍,fixture 最简单的功能就是传值的作用。什么意思呢,test_func(get_num)是一个测试函数,而 get_num 就是一个 fixture,即 fixture 的定义跟定义函数是一样的,只不过在此函数上面通过 @pytest.fixture( )装饰,get_num 这个 fixture 中直接返回一个数字 10。当然在实际应用中,其可能会经过一系列的计算处理,最后将得出的结果返回。在学习 fixture 之前,我们会把 get_nunm 认为是 test_func 的一个形参,因为 get_num 给 @pytest.fixture 声明为一个 fixture,其实就相当于测试函数 test_func(get_num)中的 test_func(get_num()),因此在测试函数中可以直接对 get_num 进行断言,期望 get_num 等于 9,这就是 fixture 的传值的使用方法。

我们可以看出,如果 get_num 没有声明为 fixture,那么断言语句肯定是错误的,因为 get_num 已经是函数对象了。而当 get_num 被声明为 fixture 时,get_num 就表示 get_num 函数的返回值了。

```
import pytest

@pytest.fixture()
def get_num():
    print("\nin get_num fixture...")
    return 10
def test_func(get_num):
    assert get_num == 9
```

通过 Pytest -s 执行结果可以看出,在测试函数中,get_num 确实不是一个函数对象,而是 get_num 函数的返回值 10。

(demo - HCIhXOHq) E:\demo> pytest - s

```
===================== test session starts =====================
platform win32 -- Python 3.7.9, pytest-7.1.3, pluggy-1.0.0
rootdir: E:\demo
plugins: assume-2.4.3, rerunfailures-10.2
collected 1 item

test_demo.py
in get_num fixture...
F

========================== FAILURES ==========================
_____ test_func _____

get_num = 10

    def test_func(get_num):
>       assert get_num == 9
E       assert 10 == 9

test_demo.py:9: AssertionError
================= short test summary info =================
FAILED test_demo.py::test_func - assert 10 == 9
==================== 1 failed in 0.07s ====================

(demo-HCIhXOHq) E:\demo>
```

上述就是 fixture 最简单应用的内容,这个功能非常适于当一个数据需要经过复杂的处理,但这个数据对于测试用例来说只是被测数据或就是入参的场景,如果没有 fixture 的应用,则只能通过定义函数的方式来实现用例,而通过 fixture 的方式则显得非常简洁。

6.2 fixture 嵌套的应用

上一节中已经体验了 fixture 传值的用法,本节继续演示 fixture 如何通过嵌套调用来传递值的用法,嵌套应用在定义 fixture 的时候,可同时调用另一个 fixture。

下面我们还是以代码进行解说。get_num1 是一个 fixture,返回值是 100,get_num2 也是一个 fixture,而在定义 get_num2 的时候,调用了 get_num1 这个 fixture,get_num2 返回的是 get_num1 的值与 10 的和,即 get_num2 返回的应该是 110,然后在测试函数 test_func 中使用了 get_num2 的 fixture,在测试用例中断言 get_num2 的期望值为 111,显然这个断言是错误的。

```python
import pytest
@pytest.fixture()
def get_num1():
    print("\nin get_num1 fixture...")
    return 100
@pytest.fixture()
def get_num2(get_num1):
    print("\nin get_num2 fixture...")
    return 10 + get_num1
def test_func(get_num2):
    assert get_num2 == 111
```

上述执行结果如下,可以看出,这里 get_num2 是 110,而不是函数对象,此外,在 get_num2 中,get_num1 传递的值同样是 100,而不是函数对象,这就是 fixture 嵌套的调用。

```
(demo-HCIhXOHq) E:\demo> pytest -s
==================== test session starts ====================
platform win32 -- Python 3.7.9, pytest-7.1.3, pluggy-1.0.0
rootdir: E:\demo
plugins: assume-2.4.3, rerunfailures-10.2
collected 1 item

test_demo.py
in get_num1 fixture...

in get_num2 fixture...
F

========================== FAILURES ==========================
_____ test_func _____

get_num2 = 110

    def test_func(get_num2):
>       assert get_num2 == 111
E       assert 110 == 111

test_demo.py:11: AssertionError
================== short test summary info ==================
FAILED test_demo.py::test_func - assert 110 == 111
==================== 1 failed in 0.06s ====================
```

(demo-HCIhXOHq) E:\demo>

fixture 嵌套的调用在处理复杂的数据是非常有用的。

6.3 在函数中调用多个 fixture

Pytest 中的 fixture 在传递值时除了可以嵌套调用,还可以同时调用多个 fixture,如我们定义了两个 fixture,get_num1 和 get_num2,其中 get_num1 返回 100,get_num2 返回 200,在测试函数 test_func 中同时调用两个 fixture,即在测试函数 test_func 的参数列表中并列将 get_num1 和 get_num2 写入,此时在测试函数 test_func 中就可以直接通过 get_num1 和 get_num2 使用两个 fixture 传递过来的值。

```
import pytest
@pytest.fixture()
def get_num1():
    print("\nin get_num1 fixture...")
    return 100
@pytest.fixture()
def get_num2():
    print("\nin get_num2 fixture...")
    return 200
def test_func(get_num1,get_num2):
    assert get_num1 == get_num2
```

很明显,get_num1 和 get_num2 的返回值是不相等的。在实际脚本开发过程中,可能需要用到多个不同业务的数据,而每个业务的数据都是通过 fixture 提供的,那么此时测试函数中直接调用多个 fixture 的功能就可以利用起来了。

```
(demo-HCIhXOHq) E:\demo> pytest -s
=================== test session starts ===================
platform win32 -- Python 3.7.9, pytest-7.1.3, pluggy-1.0.0
rootdir: E:\demo
plugins: assume-2.4.3, rerunfailures-10.2
collected 1 item

test_demo.py
in get_num1 fixture...

in get_num2 fixture...
F

========================= FAILURES =========================
```

```
_____ test_func _____

get_num1 = 100, get_num2 = 200

    def test_func(get_num1,get_num2):
>       assert get_num1 = get_num2
E       assert 100 == 200

test_demo.py:11: AssertionError
================= short test summary info =================
FAILED test_demo.py::test_func - assert 100 == 200
==================== 1 failed in 0.07s ====================

(demo-HCIhXOHq) E:\demo>
```

6.4 fixture 如何设置自动执行

本章前面几小节主要介绍了 fixture 传递值的作用,本小节介绍 fixture 作为函数自动执行的功能及其相关内容。

通过对下面代码进行简单分析,这里的断言是报错的,因为 get_name 的 fixture 传递的是一个空列表。

```
import pytest


@pytest.fixture()
def get_name():
    print("\nin get_name fixture...")
    return


def test_func(get_name):
    assert "周芷若" in get_name
```

我们通过执行结果验证上面分析的结果。

```
(demo-HCIhXOHq) E:\demo> pytest -s
==================== test session starts ====================
platform win32 -- Python 3.7.9, pytest-7.1.3, pluggy-1.0.0
rootdir: E:\demo
plugins: assume-2.4.3, rerunfailures-10.2
collected 1 item
```

```
test_demo.py
in get_name fixture...
F

========================= FAILURES =========================
_____ test_func _____

get_name = 

    def test_func(get_name):
>       assert "周芷若" in get_name
E       AssertionError: assert '周芷若' in

test_demo.py:10: AssertionError
================= short test summary info =================
FAILED test_demo.py::test_func - AssertionError: assert '...
=================1 failed in 0.09s =================

(demo-HCIhXOHq) E:\demo>
```

我们继续观察,如下一段代码中只是增加了一个 init_name 的函数定义,当然准确说是 init_name 的 fixture 定义,其在测试函数中并未调用或者引用。那么 fixture 有什么作用呢?

```python
import pytest


@pytest.fixture()
def get_name():
    print("\nin get_name fixture...")
    return


@pytest.fixture(autouse=True)
def init_name(get_name):
    print("\nin init fixture...")
    get_name.append("张无忌")
    get_name.append("张三丰")


def test_func(get_name):
    assert "周芷若" in get_name
```

我们还是先看上述执行结果,这里设置断言错误的目的是查看 get_name 此时的值,通过报错结果分析可以看出,此时 get_name 传递的值已经是['张无忌','张三丰']了,换言之,init_name 的 fixture 已经执行了,这里在定义 fixture 的时候,就将参数

autouse 设置为 True 了,这也是 fixture 自动执行的特性。

```
(demo-HCIhXOHq) E:\demo> pytest -s
==================== test session starts ====================
platform win32 -- Python 3.7.9, pytest-7.1.3, pluggy-1.0.0
rootdir: E:\demo
plugins: assume-2.4.3, rerunfailures-10.2
collected 1 item

test_demo.py
in get_name fixture...

in init fixture...
F

========================== FAILURES ==========================
_____ test_func _____

get_name = ['张无忌','张三丰']

    def test_func(get_name):
>       assert "周芷若" in get_name
E       AssertionError: assert '周芷若' in ['张无忌','张三丰']

test_demo.py:16: AssertionError
================== short test summary info ==================
FAILED test_demo.py::test_func - AssertionError: assert '...
==================== 1 failed in 0.07s ======================

(demo-HCIhXOHq) E:\demo>
```

6.5 通过 yield 实现 setup 和 teardown 的功能

通过 fixture 可以自动执行的特性很容易联想到,如果我们将初始化配置放在 fixture 中,fixture 自动执行不就可以作为 setup 来使用了吗?事实确实如此,我们看如下代码,定义了一个名为 setup_teardown_fixture 的 fixture,里面用打印了的一句话代表测试用例的初始化配置,然后在测试函数中打印了一句话代表测试内容。

```
import pytest


@pytest.fixture(autouse=True)
```

```python
def setup_teardown_fixture():
    print("in setup_teardown_fixture ...")


def test_func():
    print("in test_func ...")
    assert 1 == 1
```

执行结果如下所示,显然此时的fixture功能确实就同setup。

```
(demo-HCIhX0Hq) E:\demo> pytest -s
==================== test session starts ====================
platform win32 -- Python 3.7.9, pytest-7.1.3, pluggy-1.0.0
rootdir: E:\demo
plugins: assume-2.4.3, rerunfailures-10.2
collected 1 item

test_demo.py in setup_teardown_fixture ...
in test_func ...
.

==================== 1 passed in 0.01s ====================

(demo-HCIhX0Hq) E:\demo>
```

我们将代码修改一下,编写两个测试函数。

```python
import pytest

@pytest.fixture(autouse=True)
def setup_teardown_fixture():
    print("in setup_teardown_fixture ...")


def test_func1():
    print("in test_func1 ...")
    assert 1 == 1


def test_func2():
    print("in test_func2 ...")
    assert 1 == 1
```

再次执行结果中神奇的现象出现了,即在每个测试用例中都执行了fixture,这就是测试函数的setup,也是fixture最重要的功能,即非常灵活地实现测试框架中的setup和teardown功能。具体内容后面的章节会陆续详细展开讲解。

```
(demo-HCIhX0Hq) E:\demo> pytest -s
```

```
==================== test session starts ====================
platform win32 -- Python 3.7.9, pytest-7.1.3, pluggy-1.0.0
rootdir: E:\demo
plugins: assume-2.4.3, rerunfailures-10.2
collected 2 items

test_demo.py in setup_teardown_fixture...
in test_func1...
.in setup_teardown_fixture...
in test_func2...
.

==================== 2 passed in 0.02s ====================

(demo-HCIhX0Hq) E:\demo>
```

至此,fixture 已经实现了 setup 的功能。在自动化测试实践中,setup 和 teardown 都是必需的,即在测试用例执行完成后,要做一些环境恢复或者环境清理。那么 fixture 是如何实现 teardown 的呢？这里就要引入 yield 关键字了,fixture 通过 yield 关键字一分为二,yield 关键字之前的操作在测试用例之前执行,yield 关键字之后的操作则在测试用例执行之后执行。下面一段代码中的 fixture 使用了 yield 关键字,并且在 yield 关键字之前和之后分别打印了一句话来代替相关的操作。

```python
import pytest

@pytest.fixture(autouse=True)
def setup_teardown_fixture():
    print("in setup_teardown_fixture, before yield...")
    yield
    print("in setup_teardown_fixture, after yield...")

def test_func1():
    print("in test_func1...")
    assert 1 == 1

def test_func2():
    print("in test_func2...")
    assert 1 == 1
```

从执行结果中可以明显看出,在执行测试用例 test_func1 之前首先执行了 setup_teardown_fixture 中 yield 之前的部分,然后再执行 test_func1 的用例内容,执行完成 test_func1 之后又返回来执行 setup_teardown_fixture 中 yield 关键字之后的操作。同样,在执行 test_func2 之前首先执行了 setup_teardown_fixture 中 yield 关键字之前的内容，然后执行 test_func2 脚本的内容，执行完成 test_func2 之后又返回执行 setup_teardown_fixture 中 yield 关键字之后的内容。显然，这里 fixture 就完全作为测试用例的 setup 和 teardown，我们要做的只需要将测试用例的初始化配置放到 fixture 中 yield 关键字之前，将清理配置或恢复环境的操作放到 fixture 中 yield 之后即可。

```
(demo-HCIhXOHq) E:\demo> pytest -s
==================== test session starts ====================
platform win32 -- Python 3.7.9, pytest-7.1.3, pluggy-1.0.0
rootdir: E:\demo
plugins: assume-2.4.3, rerunfailures-10.2
collected 2 items

test_demo.py in setup_teardown_fixture, before yield...
in test_func1...
.in setup_teardown_fixture, after yield...
in setup_teardown_fixture, before yield...
in test_func2...
.in setup_teardown_fixture, after yield...


==================== 2 passed in 0.02s ====================

(demo-HCIhXOHq) E:\demo>
```

6.6　function 级别的 fixture

上一节讲述了 fixture 用来实现自动化测试的 setup 和 teardown 的功能。在自动化测试中，最常见的操作就是测试用例级别的 setup 和 teardown，以及有测试套级别的 setup 和 teardown，即一组测试用例拥有公共的初始化配置以及公共的清除配置或恢复环境操作，fixture 可以灵活地处理此方面的问题，在 fixture 定义的时候，可以通过 scope 的参数来指定当前 fixture 的作用范围，本小节主要介绍函数级别的 fixture。通过 scope 参数指定 function 来实现测试。首先看如下代码，这里在定义 fixture 的时候通过 scope 指定了 "function"。

```
import pytest

@pytest.fixture(autouse=True, scope="function")
def setup_teardown_fixture():
    print("in setup_teardown_fixture, before yield...")
    yield
```

```python
        print("in setup_teardown_fixture, after yield...")

def test_func1():
    print("in test_func1...")
    assert 1 == 1

def test_func2():
    print("in test_func2...")
    assert 1 == 1
```

从执行结果中可以看出，当 scope 设置为 function 时，在每个测试脚本之前都会执行 fixture 中 yield 之前的代码，而在每个测试脚本执行完成后，都会去执行 fixture 中 yield 关键字之后的内容。其实，对比上一节也可以发现，当 scope 不作设置的时候，默认 scope 的值就是 function，即在不设置 scope 的时候，fixture 就是测试函数级的。

```
(demo-HCIhXOHq) E:\demo> pytest -s
==================== test session starts ====================
platform win32 -- Python 3.7.9, pytest-7.1.3, pluggy-1.0.0
rootdir: E:\demo
plugins: assume-2.4.3, rerunfailures-10.2
collected 2 items

test_demo.py in setup_teardown_fixture, before yield...
in test_func1...
.in setup_teardown_fixture, after yield...
in setup_teardown_fixture, before yield...
in test_func2...
.in setup_teardown_fixture, after yield...


==================== 2 passed in 0.02s ====================

(demo-HCIhXOHq) E:\demo>
```

此外，当 scope 设置为 function 时，对测试类中的测试方法也是有效的，如下代码所示。

```python
import pytest

@pytest.fixture(autouse=True, scope="function")
def setup_teardown_fixture():
    print("in setup_teardown_fixture, before yield...")
    yield
    print("in setup_teardown_fixture, after yield...")
```

```python
class TestDemo:
    def test_func1(self):
        print("in test_func1 ...")
        assert 1 == 1

    def test_func2(self):
        print("in test_func2 ...")
        assert 1 == 1
```

同样,对于测试类中的测试方法,当 fixture 中 scope 设置为 function 时,在每个测试方法执行之前,首先执行 fixture 中 yield 之前的操作,当执行完每个测试方法后,则会执行 fixture 中 yield 关键字之后的操作。

```
(demo-HCIhXOHq) E:\demo> pytest -s
==================== test session starts ====================
platform win32 -- Python 3.7.9, pytest-7.1.3, pluggy-1.0.0
rootdir: E:\demo
plugins: assume-2.4.3, rerunfailures-10.2
collected 2 items

test_demo.py in setup_teardown_fixture, before yield ...
in test_func1 ...
.in setup_teardown_fixture, after yield ...
in setup_teardown_fixture, before yield ...
in test_func2 ...
.in setup_teardown_fixture, after yield ...


==================== 2 passed in 0.03s ====================

(demo-HCIhXOHq) E:\demo>
```

通过上面的例子可以看出,当 scope 设置为 function 时,即函数级 fixture 对测试文件中的测试函数以及测试类中的测试方法都是有效的。

6.7 class 级别的 fixture

上一节讲了测试函数级别的 fixture 相关知识,在实际自动化测试实践中,常常存在这样的场景,希望在每个测试类执行前执行公共的初始化操作,而当测试类执行完成之后,再执行公共的环境恢复操作,此时就需要 class 级别的 fixture 了,class 级别的 fixture 就是在定义的时候将 scope 参数设置为 class。

我们看下面的一段代码，即定义了一个 class 级的 fixture，同时有两个测试类 TestDemo1 和 TestDemo2，而在 fixture 中也有 yield 关键字。这里因为 fixture 被定义为 class，因此这里 yield 关键字之前的代码将在每个测试类执行之前执行，而 yield 关键字之后的代码，则会在每个测试类执行完成之后再执行。

```python
import pytest

@pytest.fixture(autouse = True, scope = "class")
def setup_teardown_fixture():
    print("in setup_teardown_fixture, before yield ...")
    yield
    print("in setup_teardown_fixture, after yield ...")

class TestDemo1:
    def test_func1(self):
        print("in test_func1 ...")
        assert 1 == 1

    def test_func2(self):
        print("in test_func2 ...")
        assert 1 == 1

class TestDemo2:
    def test_func3(self):
        print("in test_func3 ...")
        assert 1 == 1

    def test_func4(self):
        print("in test_func4 ...")
        assert 1 == 1
```

通过执行结果可以看出，此时确实 fixture 中 yield 关键字之前的内容是在每个类执行之前执行的，yield 关键字之后的内容则在每个类执行完成之后执行，而并不是在每个测试方法之前或之后执行了。class 级的 fixture 可以对每个类公共的操作进行处理。

```
(demo-HCIhX0Hq) E:\demo> pytest -s
==================== test session starts V
platform win32 -- Python 3.7.9, pytest-7.1.3, pluggy-1.0.0
rootdir: E:\demo
plugins: assume-2.4.3, rerunfailures-10.2
collected 4 items
```

```
test_demo.py in setup_teardown_fixture, before yield ...
in test_func1 ...
.in test_func2 ...
.in setup_teardown_fixture, after yield ...
in setup_teardown_fixture, before yield ...
in test_func3 ...
.in test_func4 ...
.in setup_teardown_fixture, after yield ...


=================== 4 passed in 0.02s ===================

(demo-HCIhXOHq) E:\demo>
```

6.8　module 级别的 fixture

在 Python 语言应用中，一般 module 是指一个 Python 文件，因此，从字面来理解 module 级的 fixture 就对单个 Python 文件起作用，即 fixture 中 yield 关键字之前的代码会在每个 Python 文件执行之前执行，而 fixture 中 yield 关键字之后的代码则会在每个 Python 文件执行完成后再执行。module 级的 fixture 的定义同样只需要将 scope 参数设置为 module 即可，如下代码所示。

```python
import pytest

@pytest.fixture(autouse=True, scope="module")
def setup_teardown_fixture():
    print("in setup_teardown_fixture, before yield ...")
    yield
    print("in setup_teardown_fixture, after yield ...")


class TestDemo1:
    def test_func1(self):
        print("in test_func1 ...")
        assert 1 == 1

    def test_func2(self):
        print("in test_func2 ...")
        assert 1 == 1


def test_func3():
    print("in test_func3 ...")
```

```
    assert 1 == 1

def test_func4():
    print("in test_func4 ...")
    assert 1 == 1
```

从上述执行结果可以看出,这里确实实现了在整个测试文件之前执行 fixture 中 yield 关键字之前的内容,而在整个测试文件执行完成之后,再执行 fixture 中 yield 关键字之后的内容。

```
(demo-HCIhXOHq) E:\demo> pytest -s
==================== test session starts ====================
platform win32 -- Python 3.7.9, pytest-7.1.3, pluggy-1.0.0
rootdir: E:\demo
plugins: assume-2.4.3, rerunfailures-10.2
collected 4 items

test_demo.py in setup_teardown_fixture, before yield ...
in test_func1 ...
. in test_func2 ...
. in test_func3 ...
. in test_func4 ...
. in setup_teardown_fixture, after yield ...


==================== 4 passed in 0.02s ====================

(demo-HCIhXOHq) E:\demo>
```

我们仔细观察,这里也存在一个问题,前面讲过 module 级的 fixture 可以在每个 Python 文件之前执行 yield 关键字之前的内容,那么假如有多个 Python 文件时,由于 fixture 的内容在当前文件,那么在执行其他 Python 文件之前与之后均会执行当前文件中的 fixture 吗?我们接着往下看。

我们编写如下文件目录。

```
demo/
  |----test_demo1.py
  |----test_demo2.py
```

fixture 在 test_demo1.py 中进行定义,test_demo1.py 代码如下所示。

```
import pytest

@pytest.fixture(autouse=True, scope="module")
def setup_teardown_fixture():
```

```python
        print("in setup_teardown_fixture, before yield ...")
        yield
        print("in setup_teardown_fixture, after yield ...")


class TestDemo:
    def test_func1(self):
        print("in test_demo1.TestDemo.test_func1 ...")
        assert 1 == 1

    def test_func2(self):
        print("in test_demo1.Testdemo.test_func2 ...")
        assert 1 == 1


def test_func3():
    print("in test_demo1.test_func3 ...")
    assert 1 == 1


def test_func4():
    print("in test_demo1.test_func4 ...")
    assert 1 == 1
```

我们发现,test_demo2.py 中的代码没有 fixture 的定义。

```python
import pytest


class TestDemo:
    def test_func1(self):
        print("\nin test_demo2.TestDemo.test_func1 ...")
        assert 1 == 1

    def test_func2(self):
        print("in test_demo2.TestDemo.test_func2 ...")
        assert 1 == 1


def test_func3():
    print("in test_demo2.test_func3 ...")
    assert 1 == 1


def test_func4():
    print("in test_demo2.test_func4 ...")
    assert 1 == 1
```

从执行结果中可以发现,此时在测试文件 test_demo1.py 之前和之后执行了 test_demo1.py 中的 module 级的 fixture,而在 test_demo2.py 文件之前与之后并没有执行

test_demo1.py 中定义的 module 级的 fixture。仔细想一下，这也是有道理的，如果 test_demo2.py 中没有定义 fixture，虽然 test_demo1.py 定义了 module 级的 fixture，但是在执行 test_demo2.py 的时候并不能知道 test_demo1.py 中是否定义过 fixture。

```
(demo-HCIhXOHq) E:\demo> pytest -s
=================== test session starts ===================
platform win32 -- Python 3.7.9, pytest-7.1.3, pluggy-1.0.0
rootdir: E:\demo
plugins: assume-2.4.3, rerunfailures-10.2
collected 8 items

test_demo1.py in setup_teardown_fixture, before yield ...
in test_demo1.TestDemo.test_func1 ...
.in test_demo1.Testdemo.test_func2 ...
.in test_demo1.test_func3 ...
.in test_demo1.test_func4 ...
.in setup_teardown_fixture, after yield ...

test_demo2.py
in test_demo2.TestDemo.test_func1 ...
.in test_demo2.TestDemo.test_func2 ...
.in test_demo2.test_func3 ...
.in test_demo2.test_func4 ...
.

=================== 8 passed in 0.03s ===================

(demo-HCIhXOHq) E:\demo>
```

在实际应用当中，我们总是希望将每个测试文件公共的操作提取出来作为公共的 fixture，即在每个测试文件前后都执行定义了公共操作的 fixture。这里就需要引入 conftest.py 文件了。conftest.py 文件可以理解为一个目录或者 package 的公共文件，因此从理论上分析，如果我们将 module 级别的 fixture 放入 conftest.py，那么此时 conftest.py 中的 fixture 理论上应该对当前目录下的每个测试文件都生效。

下面我们创建 conftest.py，整体目录如下所示。

```
demo/
    |---- conftest.py
    |---- test_demo1.py
    |---- test_demo2.py
```

其中，conftest.py 中的代码为 module 级的 fixture。

```
import pytest
```

```python
@pytest.fixture(autouse = True, scope = "module")
def setup_teardown_fixture():
    print("in setup_teardown_fixture, before yield ...")
    yield
    print("in setup_teardown_fixture, after yield ...")
```

此时,test_demo1.py 中的测试代码中只有测试函数了。

```python
import pytest

class TestDemo:
    def test_func1(self):
        print("in test_demo1.TestDemo.test_func1 ...")
        assert 1 == 1

    def test_func2(self):
        print("in test_demo1.Testdemo.test_func2 ...")
        assert 1 == 1

def test_func3():
    print("in test_demo1.test_func3 ...")
    assert 1 == 1

def test_func4():
    print("in test_demo1.test_func4 ...")
    assert 1 == 1
```

test_demo2.py 中的测试代码如下所示。

```python
import pytest

class TestDemo:
    def test_func1(self):
        print("\nin test_demo2.TestDemo.test_func1 ...")
        assert 1 == 1

    def test_func2(self):
        print("in test_demo2.TestDemo.test_func2 ...")
        assert 1 == 1

def test_func3():
    print("in test_demo2.test_func3 ...")
    assert 1 == 1
```

```
def test_func4():
    print("in test_demo2.test_func4 ...")
    assert 1 == 1
```

从上述执行结果可以看出,在每个测试文件即模块之前与之后均执行了 conftest.py 中 module 级的 fixture。

```
(demo-HCIhXOHq) E:\demo> pytest -s
==================== test session starts ====================
platform win32 -- Python 3.7.9, pytest-7.1.3, pluggy-1.0.0
rootdir: E:\demo
plugins: assume-2.4.3, rerunfailures-10.2
collected 8 items

test_demo1.py in setup_teardown_fixture, before yield ...
in test_demo1.TestDemo.test_func1 ...
.in test_demo1.Testdemo.test_func2 ...
.in test_demo1.test_func3 ...
.in test_demo1.test_func4 ...
.in setup_teardown_fixture, after yield ...

test_demo2.py in setup_teardown_fixture, before yield ...

in test_demo2.TestDemo.test_func1 ...
.in test_demo2.TestDemo.test_func2 ...
.in test_demo2.test_func3 ...
.in test_demo2.test_func4 ...
.in setup_teardown_fixture, after yield ...


==================== 8 passed in 0.03s ====================

(demo-HCIhXOHq) E:\demo>
```

6.9　package 级别的 fixture

Package 级别的 fixture,顾名思义就是在一个 package 执行前执行 fixture 中 yield 关键之前的代码,而在一个 package 执行完成后再执行 fixture 中 yield 关键字之后的代码。而这里的 package,简单来说就是目录,也就是说 package 级的 fixture 是控制在当前目录执行之前和执行之后进行的一些操作。

下面以一个实例来演示一下,在 demo 目录中有两个测试脚本和一个 conftest.py

文件。

```
demo
    |----conftest.py
    |----test_demo1.py
    |----test_demo2.py
```

其中 conftest.py 定义了 package 级别的 fixture，conftest.py 中的代码。

```python
import pytest

@pytest.fixture(autouse = True, scope = "package")
def setup_teardown_fixture():
    print("in conftest.setup_teardown_fixture, before yield ...")
    yield
    print("in conftest.setup_teardown_fixture, after yield ...")
```

其中 test_demo1.py 中的测试代码如下所示。

```python
import pytest


class TestDemo:
    def test_func1(self):
        print("in test_demo1.TestDemo.test_func1 ...")
        assert 1 == 1

    def test_func2(self):
        print("in test_demo1.Testdemo.test_func2 ...")
        assert 1 == 1


def test_func3():
    print("in test_demo1.test_func3 ...")
    assert 1 == 1


def test_func4():
    print("in test_demo1.test_func4 ...")
    assert 1 == 1
```

我们再来看 test_demo2.py 中的代码。

```python
import pytest


class TestDemo:
    def test_func1(self):
        print("\nin test_demo2.TestDemo.test_func1 ...")
```

```python
        assert 1 == 1

    def test_func2(self):
        print("in test_demo2.TestDemo.test_func2 ...")
        assert 1 == 1

def test_func3():
    print("in test_demo2.test_func3 ...")
    assert 1 == 1

def test_func4():
    print("in test_demo2.test_func4 ...")
    assert 1 == 1
```

这里需要注意，package 级别的 fixture 是指在当前目录执行之前与执行之后执行的，这里的当前目录是指定义 fixture 的文件所在的目录。比如这里在 conftest.py 中定义了 fixture，当前目录就是指 conftest.py 文件所在的目录，即 demo，而我们发现 test_demo1.py 和 test_demo2.py 均在 demo 目录中，因此，这里的 fixture 会在 test_demo1.py 和 test_demo2.py 执行之前与执行之后执行。

```
(demo-HCIhX0Hq) E:\demo> pytest -s
==================== test session starts ====================
platform win32 -- Python 3.7.9, pytest-7.1.3, pluggy-1.0.0
rootdir: E:\demo
plugins: assume-2.4.3, rerunfailures-10.2
collected 8 items

test_demo1.py in conftest.setup_teardown_fixture, before yield ...
in test_demo1.TestDemo.test_func1 ...
.in test_demo1.TestDemo.test_func2 ...
.in test_demo1.test_func3 ...
.in test_demo1.test_func4 ...
.
test_demo2.py
in test_demo2.TestDemo.test_func1 ...
.in test_demo2.TestDemo.test_func2 ...
.in test_demo2.test_func3 ...
.in test_demo2.test_func4 ...
.in conftest.setup_teardown_fixture, after yield ...


==================== 8 passed in 0.03s ====================
```

```
(demo-HCIhXOHq) E:\demo>
```

根据上面的解释,我们可以把 fixture 的定义直接放在 test_demo1.py 或 test_demo2.py 中,比如这里将 conftest.py 中的 fixture 注释掉,而在 test_demo1.py 中增加 fixture 注释 test_demo1.py 中的代码。

```python
import pytest

@pytest.fixture(autouse=True, scope="package")
def setup_teardown_fixture():
    print("in setup_teardown_fixture, before yield ...")
    yield
    print("in setup_teardown_fixture, after yield ...")

class TestDemo:
    def test_func1(self):
        print("in test_demo1.TestDemo.test_func1 ...")
        assert 1 == 1

    def test_func2(self):
        print("in test_demo1.Testdemo.test_func2 ...")
        assert 1 == 1

def test_func3():
    print("in test_demo1.test_func3 ...")
    assert 1 == 1

def test_func4():
    print("in test_demo1.test_func4 ...")
    assert 1 == 1
```

这里当前目录仍然是指 fixture 定义文件,即 test_demo1.py 的目录,也就是 demo,通过执行结果也很容易看出,此时 fixture 还是在 test_demo1.py 和 test_demo2.py 执行之前与执行之后才执行。

```
(demo-HCIhXOHq) E:\demo> pytest -s
==================== test session starts ====================
platform win32 -- Python 3.7.9, pytest-7.1.3, pluggy-1.0.0
rootdir: E:\demo
plugins: assume-2.4.3, rerunfailures-10.2
collected 8 items

test_demo1.py in setup_teardown_fixture, before yield ...
in test_demo1.TestDemo.test_func1 ...
```

```
.in test_demo1.Testdemo.test_func2 ...
.in test_demo1.test_func3 ...
.in test_demo1.test_func4 ...

test_demo2.py
in test_demo2.TestDemo.test_func1 ...
.in test_demo2.TestDemo.test_func2 ...
.in test_demo2.test_func3 ...
.in test_demo2.test_func4 ...
.in setup_teardown_fixture, after yield ...


===================8 passed in 0.03s ====================

(demo - HCIhXOHq) E:\demo>
```

虽然package级别的fixture在测试用例中或在当前目录下的conftest.py中定义效果都是一样的，但在实际应用中，我们仍然推荐使用在conftest.py中定义，因为此时的fixture相当于是在当前目录中所有的测试用例执行之前与之后进行的操作，在conftest.py中，可以理解为当前目录的公共操作，而如果放到某一个用例中，则很容易被忽视，而导致新用户无法理解的问题。

为了更好地理解package级别的fixture的应用情况，下面再看一个涉及多层目录即多个package时的应用场景，文件目录结构如下所示，这里有三个目录，demo目录，demo1目录，demo2目录，其中demo1和demo2在demo目录中，demo目录中还有test_demo1.py和test_demo2.py以及conftest.py文件。

```
demo/
    |----contest.py
    |----test_demo1.py
    |----test_demo2.py
    |----demo01/
            |----__init__.py
            |----conftest.py
            |----test_demo1.py
            |----test_demo2.py
    |----demo02/
            |----__init__.py
            |----conftest.py
            |----test_demo1.py
            |----test_demo2.py
```

其中，demo/conftest.py文件的代码如下所示。

```python
import pytest

@pytest.fixture(autouse = True, scope = "package")
def setup_teardown_fixture():
    print("in conftest.setup_teardown_fixture, before yield ...")
    yield
    print("in conftest.setup_teardown_fixture, after yield ...")
```

我们再看 demo/test_demo1.py 文件的代码。

```python
import pytest

class TestDemo:
    def test_func1(self):
        print("in test_demo1.TestDemo.test_func1 ...")
        assert 1 == 1

    def test_func2(self):
        print("in test_demo1.Testdemo.test_func2 ...")
        assert 1 == 1

def test_func3():
    print("in test_demo1.test_func3 ...")
    assert 1 == 1

def test_func4():
    print("in test_demo1.test_func4 ...")
    assert 1 == 1
```

接着看 demo/test_demo2.py 文件的代码。

```python
import pytest

class TestDemo:
    def test_func1(self):
        print("\nin test_demo2.TestDemo.test_func1 ...")
        assert 1 == 1

    def test_func2(self):
        print("in test_demo2.TestDemo.test_func2 ...")
        assert 1 == 1

def test_func3():
    print("in test_demo2.test_func3 ...")
    assert 1 == 1
```

```python
def test_func4():
    print("in test_demo2.test_func4...")
    assert 1 == 1
```

下面是 demo01 目录中 demo/demo01/conftest.py 文件代码。

```python
import pytest


@pytest.fixture(autouse = True, scope = "package")
def setup_teardown_fixture():
    print("demo01.conftest.in setup_teardown_fixture, before yield...")
    yield
    print("demo01.conftest.in setup_teardown_fixture, after yield...")
```

其中 demo/demo01/test_demo1.py 文件代码。

```python
import pytest


class TestDemo:
    def test_func1(self):
        print("in demo01.test_demo1.TestDemo.test_func1...")
        assert 1 == 1

    def test_func2(self):
        print("in demo01.test_demo1.Testdemo.test_func2...")
        assert 1 == 1


def test_func3():
    print("in demo01.test_demo1.test_func3...")
    assert 1 == 1


def test_func4():
    print("in demo01.test_demo1.test_func4...")
    assert 1 == 1
```

demo/demo01/test_demo2.py 文件代码。

```python
import pytest


class TestDemo:
    def test_func1(self):
        print("\nin demo01.test_demo2.TestDemo.test_func1...")
        assert 1 == 1
```

```python
    def test_func2(self):
        print("in demo01.test_demo2.TestDemo.test_func2 ...")
        assert 1 == 1


def test_func3():
    print("in demo01.test_demo2.test_func3 ...")
    assert 1 == 1


def test_func4():
    print("in demo01.test_demo2.test_func4 ...")
    assert 1 == 1
```

下面是 demo02 目录中的 demo/demo02/conftest.py 文件代码。

```python
import pytest


@pytest.fixture(autouse=True, scope="package")
def setup_teardown_fixture():
    print("in demo02.conftest.setup_teardown_fixture, before yield ...")
    yield
    print("in demo02.conftest.setup_teardown_fixture, after yield ...")
```

其中 demo/demo02/test_demo1.py 文件代码。

```python
import pytest


class TestDemo:
    def test_func1(self):
        print("in demo02.test_demo1.TestDemo.test_func1 ...")
        assert 1 == 1

    def test_func2(self):
        print("in demo02.test_demo1.Testdemo.test_func2 ...")
        assert 1 == 1


def test_func3():
    print("in demo02.test_demo1.test_func3 ...")
    assert 1 == 1


def test_func4():
    print("in demo02.test_demo1.test_func4 ...")
    assert 1 == 1
```

demo/demo02/test_demo2.py 文件的代码。

```python
import pytest

class TestDemo:
    def test_func1(self):
        print("\nin test_demo2.TestDemo.test_func1...")
        assert 1 == 1

    def test_func2(self):
        print("in test_demo2.TestDemo.test_func2...")
        assert 1 == 1

def test_func3():
    print("in test_demo2.test_func3...")
    assert 1 == 1

def test_func4():
    print("in test_demo2.test_func4...")
    assert 1 == 1
```

通过对上述代码分析，可以看到 demo 目录、demo01 目录、demo02 目录下均有 conftest.py，并且每个 conftest.py 均定义了 package 级别的 fixture，需要注意，demo/conftest.py 中定义的 package 级别的 fixture，是指 demo 目录中的所有脚本（包含 demo01 和 demo02 目录的脚本）执行之前与执行之后会执行的操作，而 demo01/conftest.py 中定义的 package 级别的 fixture 则是在 demo01 目录中所有的脚本之前与之后会执行 fixture。同理，demo02/conftest.py 文件中定义的 package 级别的 fixture 会在 demo02 目录中的所有脚本之前与之后执行。

可以看出，执行结果与上述分析的结果完全一致，至此可以看出，package 级别的 fixture 可以用于处理一个目录下所有脚本之前与之后需要进行的环境预处理以及环境恢复操作。

```
(demo-HCIhXOHq) E:\demo> pytest -s
==================== test session starts ====================
platform win32 -- Python 3.7.9, pytest-7.1.3, pluggy-1.0.0
rootdir: E:\demo
plugins: assume-2.4.3, rerunfailures-10.2
collected 24 items

test_demo1.py in conftest.setup_teardown_fixture, before yield...
in test_demo1.TestDemo.test_func1...
.in test_demo1.Testdemo.test_func2...
.in test_demo1.test_func3...
.in test_demo1.test_func4...
```

test_demo2.py
in test_demo2.TestDemo.test_func1 ...
. in test_demo2.TestDemo.test_func2 ...
. in test_demo2.test_func3 ...
. in test_demo2.test_func4 ...
.
demo01\test_demo1.py demo01.conftest.in setup_teardown_fixture, before yield ...
in demo01.test_demo1.TestDemo.test_func1 ...
. in demo01.test_demo1.Testdemo.test_func2 ...
. in demo01.test_demo1.test_func3 ...
. in demo01.test_demo1.test_func4 ...
.
demo01\test_demo2.py
in demo01.test_demo2.TestDemo.test_func1 ...
. in demo01.test_demo2.TestDemo.test_func2 ...
. in demo01.test_demo2.test_func3 ...
. in demo01.test_demo2.test_func4 ...
. demo01.conftest.in setup_teardown_fixture, after yield ...

demo02\test_demo1.py in demo02.conftest.setup_teardown_fixture, before yield ...
in demo02.test_demo1.TestDemo.test_func1 ...
. in demo02.test_demo1.Testdemo.test_func2 ...
. in demo02.test_demo1.test_func3 ...
. in demo02.test_demo1.test_func4 ...
.
demo02\test_demo2.py
in test_demo2.TestDemo.test_func1 ...
. in test_demo2.TestDemo.test_func2 ...
. in test_demo2.test_func3 ...
. in test_demo2.test_func4 ...
. in demo02.conftest.setup_teardown_fixture, after yield ...
in conftest.setup_teardown_fixture, after yield ...

==================== 24 passed in 0.09s ====================

(demo - HCIhX0Hq) E:\demo>

6.10　session 级别的 fixture

Session 即会话,这里所谓的会话就是执行 Pytest 命令的整个过程,即在执行

Pytest 命令后，首先执行的是 session 级别的 fixture 中 yield 关键字之前的代码，然后再执行所有的测试用例，在所有的测试用例执行完成之后，再执行 session 级别的 fixture 中 yield 关键字之后的代码。换句话说，session 级别的 fixture 的主要作用就是提供全局性的环境预处理和环境恢复操作。

下面就以 6.9 节中的代码为例，在 demo/conftest.py 文件中定义一个 session 级别的 fixture，即 demo/conftest.py 文件的代码，其他文件目录和文件内容保持不变。

```python
import pytest

@pytest.fixture(autouse = True, scope = "session")
def global_fixture():
    print("in conftest.global_fixture, before yield ...")
    yield
    print("in conftest.global_fixture, after yield ...")


@pytest.fixture(autouse = True, scope = "package")
def setup_teardown_fixture():
    print("in conftest.setup_teardown_fixture, before yield ...")
    yield
    print("in conftest.setup_teardown_fixture, after yield ...")
```

从上述执行结果中可以看出，在执行所有脚本之前执行了 session 级别的 fixture 中 yield 关键字之前的代码，而在所有脚本执行完成之后执行了 session 级别的 fixture 中 yield 关键字之后的代码。

```
(demo-HCIhXOHq) E:\demo> pytest -s
===================== test session starts =====================
platform win32 -- Python 3.7.9, pytest-7.2.0, pluggy-1.0.0
rootdir: E:\demo
plugins: assume-2.4.3, rerunfailures-10.2
collected 18 items

test_demo1.py in conftest.global_fixture, before yield ...
in conftest.setup_teardown_fixture, before yield ...
in test_demo1.TestDemo.test_func1 ...
.in test_demo1.Testdemo.test_func2 ...
.in test_demo1.test_func3 ...
.in test_demo1.test_func4 ...
.
test_demo2.py
in test_demo2.TestDemo.test_func1 ...
.in test_demo2.TestDemo.test_func2 ...
```

```
.in test_demo2.test_func3 ...
.in test_demo2.test_func4 ...
.
demo01\test_demo1.py demo01.conftest.in demo01_setup_teardown_fixture, before yield ...
in demo01.test_demo1.test_func4 ...
.
demo01\test_demo2.py in demo01.test_demo2.test_func4 ...
.demo01.conftest.in demo01_setup_teardown_fixture, after yield ...

demo02\test_demo1.py in demo02.conftest.demo02_setup_teardown_fixture, before yield ...
in demo02.test_demo1.TestDemo.test_func1 ...
.in demo02.test_demo1.Testdemo.test_func2 ...
.in demo02.test_demo1.test_func3 ...
.in demo02.test_demo1.test_func4 ...
.
demo02\test_demo2.py
in test_demo2.TestDemo.test_func1 ...
.in test_demo2.TestDemo.test_func2 ...
.in test_demo2.test_func3 ...
.in test_demo2.test_func4 ...
.in demo02.conftest.demo02_setup_teardown_fixture, after yield ...
in conftest.setup_teardown_fixture, after yield ...
in conftest.global_fixture, after yield ...


==================== 18 passed in 0.09s ====================


(demo - HCIhX0Hq) E:\demo>
```

至此,fixture 的所有的作用范围介绍完成了,我们总结一下,fixture 的作用范围包含 function、class、module、package、session,这个作用范围是由小到大的。如果能灵活应用这些不同作用范围的 fixture,就可以根据具体的业务需求设计出非常灵活的测试框架。当然任何事情都是非单方面的,当设计达到完美的时候就意味着代码的阅读能力下降或者说需要对 Pytest 的 fixture 非常了解才能很好地看懂代码。当然这种矛盾也是要根据具体的团队的代码编写能力水平等综合因素来决定。

6.11 fixture 的覆盖原则

前面几节非常详细地介绍了 fixture 不同的作用范围,本小节将介绍一下 fixture 的覆盖原则。在讲 fixture 的覆盖原则之前,这里首先设置出这样一个应用场景,现在有 10 个测试用例,其中 9 个测试用例前置操作和后置操作是一样的,而只有一个用例的前置操作和后置操作不一样,根据业务的划分,这 10 个用例属于测试套的。我们看这种情况下该怎么处理呢?

带着上面的问题,看下面的例子,这里为了容易理解,只使用两个文件,conftest.py 和 test_demo1.py,其中 conftest.py 中定义了一个函数级的 fixture,代码如下所示。

```
import pytest

@pytest.fixture(autouse = True, scope = "function")
def conftest_function_fixture():
    print("in conftest_function_fixture, before yield ...")
    yield
    print("in conftest_function_fixture, after yield ...")
```

这里 test_demo1.py 中也定义了一个函数级的 fixture,要注意,conftest.py 和 test_demo1.py 中的两个函数级的 fixture 的名称是不一样的。

```
import pytest

@pytest.fixture(autouse = True, scope = "function")
def test_demo1_function_fixture():
    print("in test_demo1_function_fixture, before yield ...")
    yield
    print("in test_demo1_function_fixture, after yield ...")

def test_func1():
    print("in test_demo1.test_func1 ...")
    assert 1 == 1
```

从上述执行结果中可以发现,此时两个 fixture 都执行了,而且 conftest.py 中的函数级的 fixture 首先在 test_demo1.py 中的 fixture 之前执行。

```
(demo-HCIhXOHq) E:\demo> pytest -s
==================== test session starts ====================
platform win32 -- Python 3.7.9, pytest-7.2.0, pluggy-1.0.0
rootdir: E:\demo
plugins: assume-2.4.3, rerunfailures-10.2
collected 1 item

test_demo1.py in conftest_function_fixture, before yield ...
in test_demo1_function_fixture, before yield ...
in test_demo1.test_func1 ...
.in test_demo1_function_fixture, after yield ...
in conftest_function_fixture, after yield ...
```

=================== 1 passed in 0.02s ====================

(demo-HCIhXOHq) E:\demo>

下面对 conftest.py 中的 fixture 进行修改。

```python
import pytest

@pytest.fixture(autouse=True, scope="function")
def function_fixture():
    print("in conftest_function_fixture, before yield ...")
    yield
    print("in conftest_function_fixture, after yield ...")
```

同时,test_demo1.py 中也做修改,此时,conftest.py 和 test_demo1.py 中的 fixture 的名称相同的。

```python
import pytest

@pytest.fixture(autouse=True, scope="function")
def function_fixture():
    print("in test_demo1_function_fixture, before yield ...")
    yield
    print("in test_demo1_function_fixture, after yield ...")

def test_func1():
    print("in test_demo1.test_func1 ...")
    assert 1 == 1
```

从上述执行结果中可以看出,conftest.py 中的 fixture 并未执行,此时发生了覆盖,这就是 fixture 中的就近覆盖原则,这里所说的就近是指离脚本近。

```
(demo-HCIhXOHq) E:\demo> pytest -s
=================== test session starts ====================
platform win32 -- Python 3.7.9, pytest-7.2.0, pluggy-1.0.0
rootdir: E:\demo
plugins: assume-2.4.3, rerunfailures-10.2
collected 1 item

test_demo1.py in test_demo1_function_fixture, before yield ...
in test_demo1.test_func1 ...
.in test_demo1_function_fixture, after yield ...
```

```
==================== 1 passed in 0.02s ====================
```

(demo-HCIhX0Hq) E:\demo>

下面我们继续完善,此时有三个文件,conftest.py、test_demo1.py 和 test_demo2.py,其中 conftest.py 和 test_demo1.py 的内容不变,新增的 test_demo2.py,test_demo2.py 代码中没有函数级 fixture。

```
def test_func1():
    print("in test_demo2.test_func1 ...")
    assert 1 == 1
```

仔细观察可以发现,此时在执行 test_demo1.py 中的测试用例时,执行了 test_demo1.py 中的 fixture,而未执行 conftest.py 中的 fixture,在执行 test_demo2.py 中的测试脚本时,执行了 conftest.py 中的 fixture,这是因为 test_demo1.py 中的 fixture 对 conftest.py 中的 fixture 进行了覆盖,而在执行 test_demo2.py 时,因为 test_demo2.py 中没有同名的 fixture,因此 conftest.py 中的 fixture 就不会被覆盖,因而会被执行。

```
(demo-HCIhX0Hq) E:\demo> pytest -s
==================== test session starts ====================
platform win32 -- Python 3.7.9, pytest-7.2.0, pluggy-1.0.0
rootdir: E:\demo
plugins: assume-2.4.3, rerunfailures-10.2
collected 2 items

test_demo1.py in test_demo1_function_fixture, before yield ...
in test_demo1.test_func1 ...
.in test_demo1_function_fixture, after yield ...

test_demo2.py in conftest_function_fixture, before yield ...
in test_demo2.test_func1 ...
.in conftest_function_fixture, after yield ...


==================== 2 passed in 0.03s ====================
```

(demo-HCIhX0Hq) E:\demo>

总结一下,fixture 中的覆盖原则是就近原则,即离脚本更近的同名 fixture 会覆盖外层的 fixture,但是当 fixture 名称不同时,则会全部执行,且从外往里的顺序执行,当然在用例执行完成之后的操作是从里往外的顺序,这就是典型的面向切面的编程思想。

6.12　yield 的缺陷及解决方案

通过本章前面的讲解，相信大家对 fixture 的功能也有了一定的了解，本小节将介绍 yield 关键的缺陷以及解决方案等相关内容。

首先看如下的测试代码，很明显，这里在 yield 关键字之前有一个除以 0 的错误，当脚本执行到除以 0 的代码时，会触发错误，那么此时 Pytest 又是怎么处理的呢？

```
import pytest

@pytest.fixture(autouse=True, scope="function")
def function_fixture():
    print("in test_demo1_function_fixture, before yield ...")
    a = 1/0
    yield
    print("in test_demo1_function_fixture, after yield ...")


def test_func1():
    print("in test_demo1.test_func1 ...")
    assert 1 == 1
```

我们仔细观察可以发现，这里 yield 关键字之后的代码并未执行。这就是 yield 带来的缺陷。

```
(demo-HCIhX0Hq) E:\demo> pytest -s
==================== test session starts ====================
platform win32 -- Python 3.7.9, pytest-7.2.0, pluggy-1.0.0
rootdir: E:\demo
plugins: assume-2.4.3, rerunfailures-10.2
collected 1 item

test_demo1.py in test_demo1_function_fixture, before yield ...
E

========================== ERRORS ==========================
_____ ERROR at setup of test_func1 _____

    @pytest.fixture(autouse=True, scope="function")
    def function_fixture():
        print("in test_demo1_function_fixture, before yield ...")
>       a = 1/0
E       ZeroDivisionError: division by zero
```

```
test_demo1.py:6: ZeroDivisionError
================= short test summary info =================
ERROR test_demo1.py::test_func1 - ZeroDivisionError: division by zero
==================== 1 error in 0.08s ====================
```

(demo-HCIhXOHq) E:\demo>

我们知道,在很多应用中,fixture 是被用来处理前置和后置操作的,当 yield 之前有多个操作步骤时,比如类似连接数据库、登录账号、配置环境变量等,之前由于不小心写错了代码,导致了错误,此时带来的结果就很明显。由于 yield 关键字之后的代码未被执行,此时已经配置了的环境变量等问题就需要手工去恢复,这是很麻烦的。那么怎么办呢?

有一种解决方案是通过 request.addfinalizer 方法处理,这种方法就是在执行前置操作之前,事先提醒要执行哪些后置操作,这样即使在执行前置操作时由于语法等错误失败了,Pytest 仍然只知道要执行哪些后置操作。

```
import pytest

@pytest.fixture(autouse=True, scope="function")
def function_fixture(request):
    def teardown():
        print("in test_demo1_function_fixture, after yield ...")
    request.addfinalizer(teardown)
    print("in test_demo1_function_fixture, before yield ...")
    a = 1/0
    yield


def test_func1():
    print("in test_demo1.test_func1 ...")
    assert 1 == 1
```

我们发现,此时 yield 关键之前仍然有除以 0 的操作,虽然执行到代码就报错失败了,但最后仍然把 teardown 的操作执行了。

```
(demo-HCIhXOHq) E:\demo> pytest -s
==================== test session starts ====================
platform win32 -- Python 3.7.9, pytest-7.2.0, pluggy-1.0.0
rootdir: E:\demo
plugins: assume-2.4.3, rerunfailures-10.2
collected 1 item
```

```
test_demo1.py in test_demo1_function_fixture, before yield...
Ein test_demo1_function_fixture, after yield...


========================== ERRORS ==========================
_____ ERROR at setup of test_func1 _____

request = <SubRequest 'function_fixture' for <Function test_func1>>

    @pytest.fixture(autouse=True,scope="function")
    def function_fixture(request):
        def teardown():
            print("in test_demo1_function_fixture, after yield...")
        request.addfinalizer(teardown)
        print("in test_demo1_function_fixture, before yield...")
>       a = 1/0
E       ZeroDivisionError: division by zero

test_demo1.py:9: ZeroDivisionError
================= short test summary info =================
ERROR test_demo1.py::test_func1 - ZeroDivisionError: division by zero
==================== 1 error in 0.08s ====================

(demo-HCIhX0Hq) E:\demo>
```

上面的优化方法能解决 teardown 不执行的问题，但除了不执行，还有另外的问题，比如 setup 有 5 个步骤，teardown 有对应的 5 个步骤，然而 setup 中第二个步骤中报错了，此时相当于 setup 只成功执行了一个步骤，而上述解决方案中的 teardown 中需把 5 个步骤全部执行。

为了解决上面的问题，我们让每一个 fixture 的 setup 只做原子操作，teardown 也只做与 setup 对应的原子操作，例如代码中 f1、f2、f3 中的 setup 和 teardown 分别只做原子操作，假设在 f2 的 setup 中由于某种原因出现错误，因此 f2 没有必要执行 teardown，此时，即使 f2 中 setup 失败了，f1 的 teardown 仍然执行，如此则达到了只要 setup 成功了就会执行其对应的 teardown 操作。Test_demo1.py 代码如下所示。

```
import pytest


@pytest.fixture(scope="function",autouse=True)
def f1():
    print("\nin f1 fixture setup...")
    yield
```

```
        print("\nin f1 fixture teardown...")
@pytest.fixture(scope = "function", autouse = True)
def f2(f1):
        print("\nin f2 fixture setup...")
        a = 1/0
        yield
        print("\nin f2 fixture teardown...")
@pytest.fixture(scope = "function", autouse = True)
def f3(f2):
        print("\nin f3 fixture setup...")
        yield
        print("\nin f3 fixture teardown...")
def test_func():
        print("\nin test_func...")
        assert 1 == 1
```

从上述执行结果中,可以发现,这里就可以做到了,只要前置操作执行成功,对应的后置操作就会执行,若前置操作执行失败了,则对应的后置操作不会被执行。

```
(demo-HCIhXOHq) E:\demo> pytest -s
==================== test session starts ====================
platform win32 -- Python 3.7.9, pytest-7.2.0, pluggy-1.0.0
rootdir: E:\demo
plugins: assume-2.4.3, rerunfailures-10.2
collected 1 item

test_demo1.py
in f1 fixture setup...

in f2 fixture setup...
E
in f1 fixture teardown...


========================== ERRORS ==========================
_____ ERROR at setup of test_func _____

f1 = None

    @pytest.fixture(scope = "function", autouse = True)
    def f2(f1):
        print("\nin f2 fixture setup...")
>       a = 1/0
E       ZeroDivisionError: division by zero
```

```
test_demo1.py:12: ZeroDivisionError
================= short test summary info =================
ERROR test_demo1.py::test_func - ZeroDivisionError: division by zero
=================== 1 error in 0.08s ===================
```

(demo-HCIhX0Hq) E:\demo>

上述解决方案,虽然理论上合理,但由于要拆分多个fixture,带来了代码可读性、冗余性等诸多问题,因此,在实际应用中,可以根据具体情况选择对应的解决方案处理。

第 7 章 fixture 的高级应用

前面已经介绍了 fixture 的基本应用,本章将介绍 fixture 的高级应用知识。通过对 fixture 的高级功能的灵活应用学习,可以设计灵活,功能强大的自动化测试框架。

7.1 通过 request 动态获取或配置测试脚本的属性

本小节将介绍如何通过 request 这个 fixture 来动态获取测试文件的属性或配置测试脚本的属性,在此之前,我们首先看一个简单的实例。

我们先准备两个文件,即 conftest.py 和 test_demo1.py,其中 conftest.py 定义了一个 module 级别的 fixture,并在此 fixture 中获取 module 的 name 属性和 age,在 test_demo1.py 中定义了 name 和 age 变量,写了一个测试函数。

Conftest.py 中的代码如下所示。

```
import pytest

@pytest.fixture(scope="module",autouse=True)
def function_fixture(request):
    name = getattr(request.module,"name")
    age = getattr(request.module,"age")
    print("name:",name)
    print("age:",age)
    yield
```

我们看 test_demo1.py 中的代码。

```
name = "张三"
age = 20

def test_func():
    print("\nin test_func...")
    assert 1 == 1
```

这里可以看出在 conftest.py 定义的 fixture 中可以获取测试脚本的变量属性等信息,当然这个例子只是一个用于演示学习的例子,那么我们利用这样一个特性,就可以很容易处理很多事情了。

```
(demo-HCIhXOHq) E:\demo> pytest -s
==================== test session starts ====================
platform win32 -- Python 3.7.9, pytest-7.2.0, pluggy-1.0.0
rootdir: E:\demo
plugins: assume-2.4.3, rerunfailures-10.2
collected 1 item

test_demo1.py name：张三
age：20

in test_func...
.

==================== 1 passed in 0.02s ====================

(demo-HCIhXOHq) E:\demo>
```

下面举一个例子来演示上面特性的应用，即在测试函数中编写测试用例的属性，用例名称、用例编号、用例步骤、用例期望结果等信息，然后通过在 conftest.py 中定义一个 fixture，即可实现在执行脚本时，自动将这些用例信息进行打印处理。

比如 test_demo1.py 中测试代码如下。

```python
import pytest

class Testcase:
    name = "第一个测试用例"
    id = "10001"
    steps = [
        "1.测试步骤 1：xxxxx",
        "2.测试步骤 2：yyyyy"
    ]
    expects = [
        "1.期望 xxxxx",
        "2.期望 yyyyy"
    ]

def test_func():
    print("\nin test_func...")
    assert 1 == 1
```

我们看 test_demo2.py 中的代码。

```python
import pytest
```

```python
class Testcase:
    name = "第二个测试用例"
    id = "10002"
    steps = [
        "1.测试步骤 1:aaaaa",
        "2.测试步骤 2:bbbbb"
    ]
    expects = [
        "1.期望 aaaaa",
        "2.期望 bbbbb"
    ]

def test_func():
    print("\nin test_func...")
    assert 1 == 1
```

我们看到,在测试脚本中定义了用例的属性,但在测试脚本中并未对用例属性进行任何处理,这里只需要在 conftest.py 中定义一个 fixture,在 fixture 中获取测试脚本中的用例属性,然后按照自己的需求将其打印出来,conftest.py 中的代码如下所示。

```python
import pytest

@pytest.fixture(scope="module", autouse=True)
def function_fixture(request):
    testcase = getattr(request.module, "Testcase")
    print("测试用例名称:", testcase.name)
    print("测试用例 ID:", testcase.id)
    for step in testcase.steps:
        print(step)
    for expect in testcase.expects:
        print(expect)
    yield
```

执行结果如下所示,可以看出,这里就将每个用例中的属性都打印出来了,这就是 request 的 fixture 的一个非常重要的应用。

```
(demo-HClhX0Hq) E:\demo> pytest -s
==================== test session starts ====================
platform win32 -- Python 3.7.9, pytest-7.2.0, pluggy-1.0.0
rootdir: E:\demo
plugins: assume-2.4.3, rerunfailures-10.2
collected 2 items

test_demo1.py 测试用例名称:第一个测试用例
```

```
测试用例 ID:10001
1.测试步骤 1:xxxxx
2.测试步骤 2:yyyyy
1.期望 xxxxx
2.期望 yyyyy

in test_func...
.
test_demo2.py  测试用例名称:第二个测试用例
测试用例 ID:10002
1.测试步骤 1:aaaaa
2.测试步骤 2:bbbbb
1.期望 aaaaa
2.期望 bbbbb

in test_func...
.

===================2 passed in 0.03s ===================

(demo - HCIhXOHq) E:\demo>
```

此外,我们还可以通过 fixture 给模块设置属性,典型应用如设置环境变量,即环境变量独立。

下面是在 fixture 中设置 IP 变量的值。

```
import pytest

@pytest.fixture(scope = "module",autouse = True)
def function_fixture(request):
    setattr(request.module,"IP","192.168.1.10")
    yield
```

这里定义了一个变量 IP,但是赋值为空字符串,之后会在测试函数中使用。

```
IP = ""

def test_func():
    print("\nin test_func...")
    print("IP:",IP)
    assert 1 == 1
```

可以看出,此时测试函数中已经能获取到 fixture 中设置的 IP 地址,这样一来我们就可以参考这个示例做自动化测试环境变量隔离的事情了,即一套自动化测试脚本可以完全做到环境隔离,只需要在 fixture 中修改对应的环境变量,就可做到在对应的环境上执行,从而将测试脚本与环境解耦。

```
(demo-HCIhXOHq) E:\demo> pytest -s
==================== test session starts ====================
platform win32 -- Python 3.7.9, pytest-7.2.0, pluggy-1.0.0
rootdir: E:\demo
plugins: assume-2.4.3, rerunfailures-10.2
collected 1 item

test_demo1.py
in test_func...
IP: 192.168.1.10
.

==================== 1 passed in 0.02s ====================

(demo-HCIhXOHq) E:\demo>
```

7.2　通过 request 向 fixture 传递参数

本小节将主要讲解通过 request 向 fixture 传递参数,我们先看如下一段代码,在测试函数 test_func1 中使用了@pytest.mark.name("张三")修饰,这里 mark 后面的 name 其实是自定义的,此时只需要在上面 fixture 中通过 request.node.get_closet_marker("name"),即可获取测试函数中传递的值,这里将 name 打印出来。

```python
@pytest.fixture(scope="function", autouse=True)
def function_fixture(request):
    marker = request.node.get_closest_marker("name")
    if marker:
        name = marker.args[0]
        print(name)
    yield


@pytest.mark.name("张三")
def test_func1():
    print("in test_func1...")
```

执行结果如下,可以看出,在 fixture 中可以获取测试函数中 mark 标记的 fixture 中传递的参数值。

```
(demo-HCIhXOHq) E:\demo> pytest -s
```

```
=================== test session starts ===================
platform win32 -- Python 3.7.9, pytest-7.2.0, pluggy-1.0.0
rootdir: E:\demo
plugins: assume-2.4.3, rerunfailures-10.2
collected 1 item

test_demo1.py 张三
in test_func1...
.

=================== warnings summary ===================
test_demo1.py:11
  E:\demo\test_demo1.py:11: PytestUnknownMarkWarning: Unknown pytest.mark.name - is
this a typo?  You can register custom marks to avoid this warning - for details, see https://
docs.pytest.org/en/stable/how-to/mark.html
    @pytest.mark.name("张三")

-- Docs: https://docs.pytest.org/en/stable/how-to/capture-warnings.html
============== 1 passed, 1 warning in 0.02s ===============

(demo-HCIhXOHq) E:\demo>
```

这个功能应用是很灵活的,在自动化测试框架中如果应用得当,框架的灵活性、可扩展性等都会大幅提升。举一个例子,即在自动化测试脚本中指定当前测试函数需要的环境类型,比如 host 虚拟机、docker 容器,还是不指定,那么在 conftest.py 中通过 fixture 获取这个类型时,即可作出处理。当指定环境类型为 host 时,传递给一个 IP 地址 192.168.1.10,而当环境类型为 docker 时,则传递给的 IP 地址为 172.16.1.10,其他情况,不指定时会自动提供一个 IP 地址,127.0.0.1。

conftest.py 代码如下所示。

```
import pytest


@pytest.fixture(scope="function", autouse=True)
def function_fixture(request):
    marker = request.node.get_closest_marker("env_type")
    env_type = None
    if marker is not None:
        env_type = marker.args[0]
        print(env_type)
    if env_type == "host":
        setattr(request.module, "IP", "192.168.1.10")
    elif env_type == "docker":
        setattr(request.module, "IP", "172.16.1.10")
```

```python
    else:
        setattr(request.module, "IP", "127.0.0.1")
    yield
```

test_demo1.py 中的代码如下所示。

```python
import pytest

IP = ""

@pytest.mark.env_type("host")
def test_func1():
    print("\nin test_func1...")
    print("IP:",IP)
    assert 1 == 1

@pytest.mark.env_type("docker")
def test_func2():
    print("\nin test_func2...")
    print("IP:",IP)
    assert 1 == 1

def test_func3():
    print("\nin test_func3...")
    print("IP:",IP)
    assert 1 == 1
```

可以看出，在这个例子中，因为 test_demo1.py 中每个测试函数通过 fixture 指定了环境类型，这样在 conftest.py 中的 fixture 会统一处理，每个测试函数就得到了符合自己要求的 IP 地址，这在实际应用中也是非常有用的。

```
(demo-HCIhX0Hq) E:\demo> pytest -s
==================== test session starts ====================
platform win32 -- Python 3.7.9, pytest-7.2.0, pluggy-1.0.0
rootdir: E:\demo
plugins: assume-2.4.3, rerunfailures-10.2
collected 3 items

test_demo1.py host

in test_func1...
IP: 192.168.1.10
.docker
```

```
in test_func2...
IP：172.16.1.10
.
in test_func3...
IP：127.0.0.1
.

==================== warnings summary ====================
test_demo1.py:5
  E:\demo\test_demo1.py:5: PytestUnknownMarkWarning: Unknown pytest.mark.env_type - is
this a typo?  You can register custom marks to avoid this warning - for details, see https://
docs.pytest.org/en/stable/how-to/mark.html
    @pytest.mark.env_type("host")

test_demo1.py:11
  E:\demo\test_demo1.py:11: PytestUnknownMarkWarning: Unknown pytest.mark.env_type -
is this a typo?  You can register custom marks to avoid this warning - for details, see ht-
tps://docs.pytest.org/en/stable/how-to/mark.html
    @pytest.mark.env_type("docker")

-- Docs: https://docs.pytest.org/en/stable/how-to/capture-warnings.html
=============== 3 passed, 2 warnings in 0.03s ===============

(demo-HCIhXOHq) E:\demo>
```

7.3 fixture 如何实现参数化，即数据驱动

在介绍 fixture 实现参数化之前，首先来看下 fixture 是如何传值给测试函数的，上一章我们已经介绍过 fixture 传值的用法，那么下面将要介绍 fixture 在实现前置操作和后置操作的同时向测试函数传递值的方法。

前面我们已经介绍过关键字 yield 的作用，即 yield 之前的操作在测试用例之前执行，yield 之后的操作在测试用例之后执行，yield 关键字独占一行，后面也未跟任何参数。其实 yield 后面是可以跟一个数值或者变量的，而这个数值或者变量可以传递给测试函数，即在执行完前置操作后，向测试函数传递了一个值 10，当然，此时 fixture 的名称 function_fixture 需要作为测试函数的一个形参。

```
import pytest

@pytest.fixture()
def function_fixture(request):
```

```python
    print("in function_fixture before yield ...")
    yield 10
    print("in function_fixture after yield ...")

def test_func1(function_fixture):
    print("in test_func1 ...")
    print("function_fixture",function_fixture)
    assert function_fixture == 10
```

仔细观察,可以发现在测试函数中确实获取了 fixture 传递过来的数值 10,这就是 fixture 在提供前置操作和后置操作的同时又可以向测试函数传递值的用法。

```
(demo-HCIhX0Hq) E:\demo> pytest -s
==================== test session starts ====================
platform win32 -- Python 3.7.9, pytest-7.2.0, pluggy-1.0.0
rootdir: E:\demo
plugins: assume-2.4.3, rerunfailures-10.2
collected 1 item

test_demo1.py in function_fixture before yield ...
in test_func1 ...
function_fixture 10
.in function_fixture after yield ...


==================== 1 passed in 0.02s ====================

(demo-HCIhX0Hq) E:\demo>
```

至此,了解了 yield 的传递值用法后,下面我们介绍 fixture 如何实现参数化即数据驱动。我们知道,在测试领域或自动化测试领域,数据驱动测试是一种非常重要的测试方法,主要用于接口测试、数值合法性校验等场景,原理就是对同一套逻辑提供多个不同的数值,进而将每个数值都作为一个测试用例去执行。

我们在 fixture 中让 params 提供一组数据,然后通过 yield 将 request.param 传递给测试函数。下面测试函数的功能是判断数据是否为偶数,fixture 传递了 6 个数,三个奇数三个偶数。

```python
import pytest

@pytest.fixture(scope="function",params=[1,2,3,4,5,6])
def function_fixture(request):
    print("in function_fixture before yield ...")
    yield request.param
```

```python
    print("in function_fixture after yield ...")

def test_func1(function_fixture):
    print("in test_func1 ...")
    print("function_fixture",function_fixture)
    assert function_fixture % 2 == 0
```

这里执行结果中有 6 个测试用例,其中 3 个通过,3 个失败,与上面的代码分析结果完全一致,这就是 fixture 实现参数化数据驱动测试的用法。

```
(demo-HCIhXOHq) E:\demo> pytest -s
==================== test session starts ====================
platform win32 -- Python 3.7.9, pytest-7.2.0, pluggy-1.0.0
rootdir: E:\demo
plugins: assume-2.4.3, rerunfailures-10.2
collected 6 items


test_demo1.py in function_fixture before yield ...
in test_func1 ...
function_fixture 1
Fin function_fixture after yield ...
in function_fixture before yield ...
in test_func1 ...
function_fixture 2
. in function_fixture after yield ...
in function_fixture before yield ...
in test_func1 ...
function_fixture 3
Fin function_fixture after yield ...
in function_fixture before yield ...
in test_func1 ...
function_fixture 4
. in function_fixture after yield ...
in function_fixture before yield ...
in test_func1 ...
function_fixture 5
Fin function_fixture after yield ...
in function_fixture before yield ...
in test_func1 ...
function_fixture 6
. in function_fixture after yield ...


========================= FAILURES =========================
```

_____ test_func1[1] _____

function_fixture = 1

```
    def test_func1(function_fixture):
        print("in test_func1 ...")
        print("function_fixture",function_fixture)
>       assert function_fixture % 2 == 0
E       assert (1 % 2) == 0
```

test_demo1.py:12: AssertionError
_____ test_func1[3] _____

function_fixture = 3

```
    def test_func1(function_fixture):
        print("in test_func1 ...")
        print("function_fixture",function_fixture)
>       assert function_fixture % 2 == 0
E       assert (3 % 2) == 0
```

test_demo1.py:12: AssertionError
_____ test_func1[5] _____

function_fixture = 5

```
    def test_func1(function_fixture):
        print("in test_func1 ...")
        print("function_fixture",function_fixture)
>       assert function_fixture % 2 == 0
E       assert (5 % 2) == 0
```

test_demo1.py:12: AssertionError
================ short test summary info ================
FAILED test_demo1.py::test_func1[1] - assert (1 % 2) == 0
FAILED test_demo1.py::test_func1[3] - assert (3 % 2) == 0
FAILED test_demo1.py::test_func1[5] - assert (5 % 2) == 0
================ 3 failed, 3 passed in 0.10s ================

(demo-HCIhX0Hq) E:\demo>

7.4 fixture 参数化指定用例 id

上一节我们介绍了 fixture 可以实现参数化功能的情况,下面介绍通过 Pytest-collect-only 来查看测试用例的 id,如下所示。

```python
import pytest

@pytest.fixture(scope = "function", params = [1,2,3,4,5,6])
def function_fixture(request):
    print("in function_fixture before yield...")
    yield request.param
    print("in function_fixture after yield...")

def test_func1(function_fixture):
    print("in test_func1...")
    print("function_fixture", function_fixture)
    assert function_fixture % 2 == 0
```

这里测试用例的 id 为方括号中的数字,即这里的数字是 1~6。

```
(demo-HCIhXOHq) E:\demo> pytest --collect-only
=================== test session starts ===================
platform win32 -- Python 3.7.9, pytest-7.2.0, pluggy-1.0.0
rootdir: E:\demo
plugins: assume-2.4.3, rerunfailures-10.2
collected 6 items

<Package demo>
  <Module test_demo1.py>
    <Function test_func1[1]>
    <Function test_func1[2]>
    <Function test_func1[3]>
    <Function test_func1[4]>
    <Function test_func1[5]>
    <Function test_func1[6]>

=============== 6 tests collected in 0.01s ================

(demo-HCIhXOHq) E:\demo>
```

其实在实现参数化的过程中,fixture 是支持指定测试用例 id 的,只需要在 fixture

定义的时候增加一个 ids 的列表参数即可,其中元素与参数化中的数据个数要一致,比如如下代码。

```python
import pytest

@pytest.fixture(scope="function",params=[1,2,3,4,5,6],ids=["test_01","test_02","test_03","test_04","test_05","test_06"])
def function_fixture(request):
    print("in function_fixture before yield...")
    yield request.param
    print("in function_fixture after yield...")

def test_func1(function_fixture):
    print("in test_func1...")
    print("function_fixture",function_fixture)
    assert function_fixture % 2 == 0
```

此时,pytest --collect-only 再一次通过,方括号中的测试,用例 id 已经发生了变化。

```
(demo-HCIhXOHq) E:\demo> pytest --collect-only
==================== test session starts ====================
platform win32 -- Python 3.7.9, pytest-7.2.0, pluggy-1.0.0
rootdir: E:\demo
plugins: assume-2.4.3, rerunfailures-10.2
collected 6 items

<Package demo>
  <Module test_demo1.py>
    <Function test_func1[test_01]>
    <Function test_func1[test_02]>
    <Function test_func1[test_03]>
    <Function test_func1[test_04]>
    <Function test_func1[test_05]>
    <Function test_func1[test_06]>

=============== 6 tests collected in 0.04s ================

(demo-HCIhXOHq) E:\demo>
```

7.5　fixture 参数化中指定参数使用 skip 标记

fixture 实现参数的应用,还可以通过 skip 指定数据进行跳过来完成。在前面我们已经接触过 skip 的应用了，对于一些功能尚未实现的场景，我们在编写自动化脚本的时候可以先使用 skip 跳过，那么在 fixture 实现参数的过程中，同样可以对指定的数据进行 skip 的。具体用法是，在参数化数据的列表中，对要暂时忽略的数据使用 pytest.param 进行指定 marks 即可。

```python
import pytest

@pytest.fixture(scope="function",params=[1,2,pytest.param(3,marks=pytest.mark.skip),4,pytest.param(5,marks=pytest.mark.skip),6],ids=["test_01","test_02","test_03","test_04","test_05","test_06"])
def function_fixture(request):
    print("in function_fixture before yield ...")
    yield request.param
    print("in function_fixture after yield ...")

def test_func1(function_fixture):
    print("in test_func1 ...")
    print("function_fixture",function_fixture)
    assert function_fixture % 2 == 0
```

从执行结果中可以看到，此时对于参数 3 和 5，执行结果已经标记为 skip 状态。

```
(demo-HCIhXOHq) E:\demo> pytest -s
==================test session starts ==================
platform win32 -- Python 3.7.9, pytest-7.2.0, pluggy-1.0.0
rootdir: E:\demo
plugins: assume-2.4.3, rerunfailures-10.2
collected 6 items

test_demo1.py in function_fixture before yield ...
in test_func1 ...
function_fixture 1
Fin function_fixture after yield ...
in function_fixture before yield ...
in test_func1 ...
function_fixture 2
.in function_fixture after yield ...
sin function_fixture before yield ...
in test_func1 ...
function_fixture 4
.in function_fixture after yield ...
sin function_fixture before yield ...
in test_func1 ...
function_fixture 6
```

```
    .in function_fixture after yield ...


    ========================= FAILURES =========================
    _____ test_func1[test_01] _____

function_fixture = 1

    def test_func1(function_fixture):
        print("in test_func1 ...")
        print("function_fixture",function_fixture)
>       assert function_fixture % 2 == 0
E       assert (1 % 2) == 0

test_demo1.py:12: AssertionError
================== short test summary info ==================
FAILED test_demo1.py::test_func1[test_01] - assert (1 % 2) == 0
========== 1 failed, 3 passed, 2 skipped in 0.08s ==========


(demo-HCIhXOHq) E:\demo>
```

7.6 fixture 参数化时,指定参数使用 xfail 标记

在 fixture 实现参数化的场景中,通过 xfail 标记失败的用法与 skip 的用法几乎完全一样,只需要在参数化列表中对需要标记为 xfail 的数据使用 Pytest.param 进行指定 marks 即可,代码如下所示。

```
import pytest

@pytest.fixture(scope="function",params=[1,2,pytest.param(3,marks=pytest.mark.xfail),4,pytest.param(5,marks=pytest.mark.xfail),6],ids=["test_01","test_02","test_03","test_04","test_05","test_06"])
def function_fixture(request):
    print("in function_fixture before yield ...")
    yield request.param
    print("in function_fixture after yield ...")

def test_func1(function_fixture):
    print("in test_func1 ...")
    print("function_fixture",function_fixture)
    assert function_fixture % 2 == 0
```

可以看出,此时,参数 3 和 5 均已标记为 xfail 状态。

```
(demo-HCIhX0Hq) E:\demo> pytest -s
===================== test session starts =====================
platform win32 -- Python 3.7.9, pytest-7.2.0, pluggy-1.0.0
rootdir: E:\demo
plugins: assume-2.4.3, rerunfailures-10.2
collected 6 items

test_demo1.py in function_fixture before yield ...
in test_func1 ...
function_fixture 1
Fin function_fixture after yield ...
in function_fixture before yield ...
in test_func1 ...
function_fixture 2
.in function_fixture after yield ...
in function_fixture before yield ...
in test_func1 ...
function_fixture 3
xin function_fixture after yield ...
in function_fixture before yield ...
in test_func1 ...
function_fixture 4
.in function_fixture after yield ...
in function_fixture before yield ...
in test_func1 ...
function_fixture 5
xin function_fixture after yield ...
in function_fixture before yield ...
in test_func1 ...
function_fixture 6
.in function_fixture after yield ...


========================== FAILURES ==========================
_____ test_func1[test_01] _____

function_fixture = 1

    def test_func1(function_fixture):
        print("in test_func1 ...")
        print("function_fixture",function_fixture)
```

```
>       assert function_fixture % 2 == 0
E       assert (1 % 2) == 0

test_demo1.py:12: AssertionError
================== short test summary info ==================
FAILED test_demo1.py::test_func1[test_01] - assert (1 % 2) == 0
========== 1 failed, 3 passed, 2 xfailed in 0.09s ==========

(demo-HCIhXOHq) E:\demo>
```

7.7　fixture 参数化可实现两组数据的全排列组合测试

fixture 的参数化还可以实现两组数据的全排列组合测试，比如将第一组数据 1 和 2，第二组数据 10 和 20，进行全排列为四组(1,10,),(1,20),(2,10),(2,20)测试。如要判断第一个数和第二个数是否相等，可通过 fixture 实现，只需要定义两个 fixture，每个 fixture 声明一组参数化，然后在测试函数中将这两个参数全部作为测试函数的形参即可。

```
import pytest

@pytest.fixture(scope="function",params=[1,2])
def function_fixture1(request):
    print("in function_fixture1 before yield ...")
    yield request.param
    print("in function_fixture1 after yield ...")


@pytest.fixture(scope="function",params=[10,20])
def function_fixture2(request):
    print("in function_fixture2 before yield ...")
    yield request.param
    print("in function_fixture2 after yield ...")


def test_func1(function_fixture1,function_fixture2):
    print("in test_func1 ...")
    assert function_fixture1 == function_fixture2
```

执行结果如下所示，我们看到，此时确实将两组数据进行了全排列组合测试。

```
(demo-HCIhXOHq) E:\demo> pytest -s
==================== test session starts ====================
platform win32 -- Python 3.7.9, pytest-7.2.0, pluggy-1.0.0
rootdir: E:\demo
```

plugins: assume-2.4.3, rerunfailures-10.2
collected 4 items

test_demo1.py in function_fixture1 before yield ...
in function_fixture2 before yield ...
in test_func1 ...
Fin function_fixture2 after yield ...
in function_fixture1 after yield ...
in function_fixture1 before yield ...
in function_fixture2 before yield ...
in test_func1 ...
Fin function_fixture2 after yield ...
in function_fixture1 after yield ...
in function_fixture1 before yield ...
in function_fixture2 before yield ...
in test_func1 ...
Fin function_fixture2 after yield ...
in function_fixture1 after yield ...
in function_fixture1 before yield ...
in function_fixture2 before yield ...
in test_func1 ...
Fin function_fixture2 after yield ...
in function_fixture1 after yield ...

========================= FAILURES =========================
_____ test_func1[1-10] _____

function_fixture1 = 1, function_fixture2 = 10

        def test_func1(function_fixture1,function_fixture2):
            print("in test_func1 ...")
>   assert function_fixture1 == function_fixture2
E   assert 1 == 10

test_demo1.py:17: AssertionError
_____ test_func1[1-20] _____

function_fixture1 = 1, function_fixture2 = 20

        def test_func1(function_fixture1,function_fixture2):
            print("in test_func1 ...")
>   assert function_fixture1 == function_fixture2

```
E       assert 1 == 20

test_demo1.py:17: AssertionError
_____ test_func1[2-10] _____

function_fixture1 = 2, function_fixture2 = 10

    def test_func1(function_fixture1,function_fixture2):
        print("in test_func1 ...")
>       assert function_fixture1 == function_fixture2
E       assert 2 == 10

test_demo1.py:17: AssertionError
_____ test_func1[2-20] _____

function_fixture1 = 2, function_fixture2 = 20

    def test_func1(function_fixture1,function_fixture2):
        print("in test_func1 ...")
>       assert function_fixture1 == function_fixture2
E       assert 2 == 20

test_demo1.py:17: AssertionError
================ short test summary info ================
FAILED test_demo1.py::test_func1[1-10] - assert1 == 10
FAILED test_demo1.py::test_func1[1-20] - assert 1 == 20
FAILED test_demo1.py::test_func1[2-10] - assert 2 == 10
FAILED test_demo1.py::test_func1[2-20] - assert 2 == 20
==================== 4 failed in 0.11s ====================

(demo-HCIhXOHq) E:\demo>
```

此时很容易联想到，如果三组数据实现全排列，即编写三个 fixture 会如何？这里就不再详细解释。

```
import pytest

@pytest.fixture(scope="function",params=[1,2])
def function_fixture1(request):
    print("in function_fixture1 before yield ...")
    yield request.param
    print("in function_fixture1 after yield ...")
```

```python
@pytest.fixture(scope="function", params=[10,20])
def function_fixture2(request):
    print("in function_fixture2 before yield ...")
    yield request.param
    print("in function_fixture2 after yield ...")


@pytest.fixture(scope="function", params=[100,200])
def function_fixture3(request):
    print("in function_fixture3 before yield ...")
    yield request.param
    print("in function_fixture3 after yield ...")


def test_func1(function_fixture1, function_fixture2, function_fixture3):
    print("in test_func1 ...")
    assert function_fixture1 + function_fixture2 > function_fixture3
```

执行结果如下所示,可以发现,此时共 2×2×2=8 个测试用例。

```
(demo-HCIhX0Hq) E:\demo> pytest -s
==================== test session starts ====================
platform win32 -- Python 3.7.9, pytest-7.2.0, pluggy-1.0.0
rootdir: E:\demo
plugins: assume-2.4.3, rerunfailures-10.2
collected 8 items

test_demo1.py in function_fixture1 before yield ...
in function_fixture2 before yield ...
in function_fixture3 before yield ...
in test_func1 ...
Fin function_fixture3 after yield ...
in function_fixture2 after yield ...
in function_fixture1 after yield ...
in function_fixture1 before yield ...
in function_fixture2 before yield ...
in function_fixture3 before yield ...
in test_func1 ...
Fin function_fixture3 after yield ...
in function_fixture2 after yield ...
in function_fixture1 after yield ...
in function_fixture1 before yield ...
in function_fixture2 before yield ...
in function_fixture3 before yield ...
in test_func1 ...
Fin function_fixture3 after yield ...
```

```
in function_fixture2 after yield ...
in function_fixture1 after yield ...
in function_fixture1 before yield ...
in function_fixture2 before yield ...
in function_fixture3 before yield ...
in test_func1 ...
Fin function_fixture3 after yield ...
in function_fixture2 after yield ...
in function_fixture1 after yield ...
in function_fixture1 before yield ...
in function_fixture2 before yield ...
in function_fixture3 before yield ...
in test_func1 ...
Fin function_fixture3 after yield ...
in function_fixture2 after yield ...
in function_fixture1 after yield ...
in function_fixture1 before yield ...
in function_fixture2 before yield ...
in function_fixture3 before yield ...
in test_func1 ...
Fin function_fixture3 after yield ...
in function_fixture2 after yield ...
in function_fixture1 after yield ...
in function_fixture1 before yield ...
in function_fixture2 before yield ...
in function_fixture3 before yield ...
in test_func1 ...
Fin function_fixture3 after yield ...
in function_fixture2 after yield ...
in function_fixture1 after yield ...
in function_fixture1 before yield ...
in function_fixture2 before yield ...
in function_fixture3 before yield ...
in test_func1 ...
Fin function_fixture3 after yield ...
in function_fixture2 after yield ...
in function_fixture1 after yield ...

========================= FAILURES =========================
_____ test_func1[1 - 10 - 100] _____

function_fixture1 = 1, function_fixture2 = 10
```

```
function_fixture3 = 100

    def test_func1(function_fixture1,function_fixture2,function_fixture3):
        print("in test_func1 ...")
>       assert function_fixture1 + function_fixture2 > function_fixture3
E       assert (1 + 10) > 100

test_demo1.py:23: AssertionError
_____ test_func1[1-10-200] _____

function_fixture1 = 1, function_fixture2 = 10
function_fixture3 = 200

    def test_func1(function_fixture1,function_fixture2,function_fixture3):
        print("in test_func1 ...")
>       assert function_fixture1 + function_fixture2 > function_fixture3
E       assert (1 + 10) > 200

test_demo1.py:23: AssertionError
_____ test_func1[1-20-100] _____

function_fixture1 = 1, function_fixture2 = 20
function_fixture3 = 100

    def test_func1(function_fixture1,function_fixture2,function_fixture3):
        print("in test_func1 ...")
>       assert function_fixture1 + function_fixture2 > function_fixture3
E       assert (1 + 20) > 100

test_demo1.py:23: AssertionError
_____ test_func1[1-20-200] _____

function_fixture1 = 1, function_fixture2 = 20
function_fixture3 = 200

    def test_func1(function_fixture1,function_fixture2,function_fixture3):
        print("in test_func1 ...")
>       assert function_fixture1 + function_fixture2 > function_fixture3
E       assert (1 + 20) > 200

test_demo1.py:23: AssertionError
_____ test_func1[2-10-100] _____
```

```
function_fixture1 = 2, function_fixture2 = 10
function_fixture3 = 100

    def test_func1(function_fixture1,function_fixture2,function_fixture3):
        print("in test_func1 ...")
>       assert function_fixture1 + function_fixture2 > function_fixture3
E       assert (2 + 10) > 100

test_demo1.py:23: AssertionError
_____ test_func1[2 - 10 - 200] _____

function_fixture1 = 2, function_fixture2 = 10
function_fixture3 = 200

    def test_func1(function_fixture1,function_fixture2,function_fixture3):
        print("in test_func1 ...")
>       assert function_fixture1 + function_fixture2 > function_fixture3
E       assert (2 + 10) > 200

test_demo1.py:23: AssertionError
_____ test_func1[2 - 20 - 100] _____

function_fixture1 = 2, function_fixture2 = 20
function_fixture3 = 100

    def test_func1(function_fixture1,function_fixture2,function_fixture3):
        print("in test_func1 ...")
>       assert function_fixture1 + function_fixture2 > function_fixture3
E       assert (2 + 20) > 100

test_demo1.py:23: AssertionError
_____ test_func1[2 - 20 - 200] _____

function_fixture1 = 2, function_fixture2 = 20
function_fixture3 = 200

    def test_func1(function_fixture1,function_fixture2,function_fixture3):
        print("in test_func1 ...")
>       assert function_fixture1 + function_fixture2 > function_fixture3
E       assert (2 + 20) > 200

test_demo1.py:23: AssertionError
```

```
================= short test summary info ==================
FAILED test_demo1.py::test_func1[1-10-100] - assert (1 + 10) > 100
FAILED test_demo1.py::test_func1[1-10-200] - assert (1 + 10) > 200
FAILED test_demo1.py::test_func1[1-20-100] - assert (1 + 20) > 100
FAILED test_demo1.py::test_func1[1-20-200] - assert (1 + 20) > 200
FAILED test_demo1.py::test_func1[2-10-100] - assert (2 + 10) > 100
FAILED test_demo1.py::test_func1[2-10-200] - assert (2 + 10) > 200
FAILED test_demo1.py::test_func1[2-20-100] - assert (2 + 20) > 100
FAILED test_demo1.py::test_func1[2-20-200] - assert (2 + 20) > 200
================== 8 failed in 0.16s ==================

(demo-HCIhX0Hq) E:\demo>
```

7.8 通过 usefixtures 为一个测试类调用 fixture

在介绍 userfixtures 之前，首先看一段测试代码，这里有测试类，有测试函数，有一个 fixture，如果此时想将 function 级别的 fixture 作用于域测试类中的每个测试方法，且对测试函数 test_func3 不生效，则不能在 fixture 中使用 autouse=True 的方式，因为这种方式将对测试类中的测试方法以及测试文件中的测试函数均生效。

```python
import pytest

@pytest.fixture(scope="function")
def function_fixture():
    print("in function_fixture before yield...")
    yield
    print("in function_fixture after yield...")

class TestDemo(object):
    def test_func1(self):
        print("in TestDemo.test_func1...")
        assert 1 == 1

    def test_func2(self):
        print("in TestDemo.test_func1...")
        assert 1 == 1

def test_func3():
    print("test_func3...")
    assert 1 == 1
```

当然，也可以在测试中的每个测试方法中的形参列表中写入 fixture，比如如下代码。

```python
import pytest

@pytest.fixture(scope="function")
def function_fixture():
    print("in function_fixture before yield...")
    yield
    print("in function_fixture after yield...")

class TestDemo(object):
    def test_func1(self,function_fixture):
        print("in TestDemo.test_func1...")
        assert 1 == 1

    def test_func2(self,function_fixture):
        print("in TestDemo.test_func1...")
        assert 1 == 1

def test_func3():
    print("test_func3...")
    assert 1 == 1
```

可以看出，fixtule 此时已经对测试类中的测试方法生效了，但是对测试文件中的测试函数 test_func3 是无效的。

```
(demo-HCIhXOHq) E:\demo> pytest -s
==================== test session starts ====================
platform win32 -- Python 3.7.9, pytest-7.2.0, pluggy-1.0.0
rootdir: E:\demo
plugins: assume-2.4.3, rerunfailures-10.2
collected 3 items

test_demo1.py in function_fixture before yield...
in TestDemo.test_func1...
.in function_fixture after yield...
in function_fixture before yield...
in TestDemo.test_func1...
.in function_fixture after yield...
test_func3...
.
==================== 3 passed in 0.02s ====================

(demo-HCIhXOHq) E:\demo>
```

但是这样做有一个弊端，如果测试类中有一百个测试方法，那么就需要在这 100 个测试方法的形参列表中全部列出 fixture，这样做是很麻烦的，这里就可以用 usefixtures

方法做处理。usefixture 的使用方式简单得多,只需要在测试类上面使用 @pytest.mark.usefixtures("xxx"),然后测试类中的每个测试方法即可自动使用此 fixture 了。我们看如下测试代码。

```python
import pytest

@pytest.fixture(scope = "function")
def function_fixture():
    print("in function_fixture before yield...")
    yield
    print("in function_fixture after yield...")

@pytest.mark.usefixtures("function_fixture")
class TestDemo(object):
    def test_func1(self):
        print("in TestDemo.test_func1...")
        assert 1 == 1

    def test_func2(self):
        print("in TestDemo.test_func1...")
        assert 1 == 1

def test_func3():
    print("test_func3...")
    assert 1 == 1
```

执行结果如下,显然使用 usefixtures 方法达到了较好的处理效果。

```
(demo-HCIhXOHq) E:\demo> pytest -s
==================== test session starts ====================
platform win32 -- Python 3.7.9, pytest-7.2.0, pluggy-1.0.0
rootdir: E:\demo
plugins: assume-2.4.3, rerunfailures-10.2
collected 3 items

test_demo1.py in function_fixture before yield...
in TestDemo.test_func1...
.in function_fixture after yield...
in function_fixture before yield...
in TestDemo.test_func1...
.in function_fixture after yield...
test_func3...
.

==================== 3 passed in 0.03s ====================

(demo-HCIhXOHq) E:\demo>
```

第 8 章 常见内置 fixture 的应用

8.1 如何进行文档测试

在介绍文档测试之前,我们先设置好 Python 交互解释的环境,然后输入一些参数计算等,都是一些非常简单的 Python 交互运算,相信大家在刚刚开始学 Python 语言的时候,经常练习此类运算。

```
>>> 1+1
2
>>> "hello" in "hello world"
True
>>> "hello" not in "hello world"
False
>>> 110 * 10
1100
>>> "key" in {"key":"hhh"}
True
>>>
```

Pytest 同样支持上述测试,我们将下面内容存入 test_demo.txt 文件中。大家注意一下 Pytest 支持 test*.txt 格式的本文内容。

```
>>> 1+1
2
>>> "hello" in "hello world"
True
>>> "hello" not in "hello world"
False
>>> 110 * 10
1100
>>> "key" in {"key":"hhh"}
True
```

可以发现,此时 Pytest 把原本 Python 语言的交互式文本作为一个测试用例执行了。

```
(demo-HCIhXOHq) E:\demo> pytest -s
==================== test session starts ====================
platform win32 -- Python 3.7.9, pytest-7.2.0, pluggy-1.0.0
rootdir: E:\demo
plugins: assume-2.4.3, rerunfailures-10.2
collected 1 item

test_demo.txt .

==================== 1 passed in 0.02s ====================

(demo-HCIhXOHq) E:\demo>
```

那么 Pytest 是否真正执行了呢？我们来修改一下，只修改一处错误，即将 110×10 的交互值修改为 0。在文本测试中，其实交互值即为期望值，因此这里测试用例 110× 10 期望结果为 0，从代码分析得知，这是错误的。

```
>>> 1 + 1
2
>>> "hello" in "hello world"
True
>>> "hello" not in "hello world"
False
>>> 110 * 10
0
>>> "key" in {"key":"hhh"}
True
```

可以发现，这里确实执行，且打印出了期望值 0，得到了 1 100。

```
(demo-HCIhXOHq) E:\demo> pytest -s
==================== test session starts ====================
platform win32 -- Python 3.7.9, pytest-7.2.0, pluggy-1.0.0
rootdir: E:\demo
plugins: assume-2.4.3, rerunfailures-10.2
collected 1 item

test_demo.txt F

========================== FAILURES ==========================
_____ [doctest] test_demo.txt _____
001 >>> 1 + 1
002 2
003 >>> "hello" in "hello world"
```

```
004 True
005 >>> "hello" not in "hello world"
006 False
007 >>> 110 * 10
Expected:
    0
Got:
    1100

E:\demo\test_demo.txt:7: DocTestFailure
================= short test summary info =================
FAILED test_demo.txt::test_demo.txt
==================== 1 failed in 0.03s ====================

(demo-HCIhX0Hq) E:\demo>
```

上述内容就是简单的文本测试情况，它本质是调用内置的 doctest 的 fixture。除此以外，我们还可以执行 Python 文件中的注释内容，文件名命名规则没有要求，只要是 Python 文件即可，这里定义了一个 add 函数，用于计算给定的两个数的和，那么对于这种比较简单的函数，则可以在注释中使用类似交互的方式写测试用例。比如这里写了两个用例，一个是传入 10 和 20，期望结果为 30，第二个是传入 100 和 200，期望结果为 10，从代码分析可以看出第二个期望是错误的。

```
def add(a,b):
    """
    >>> add(10,20)
    30
    >>> add(100,200)
    10
    :param a:
    :param b:
    :return:
    """
    return a + b
```

这里需要注意，执行的时候需要增加 --doctest-modules 参数，我们发现确实执行了，而且显示用例执行失败，期望结果为 10，得到了结果 300。这种方式的文本测试在 Python 代码开发中是非常有用的，比如在 Python 语言项目中，完全可以采用这种文本测试的方式进行单元测试的开发，因为单元测试本身就是白盒测试，用这种方式非常简单易用。

```
(demo-HCIhX0Hq) E:\demo> pytest --doctest-modules
==================== test session starts ====================
```

```
platform win32 -- Python 3.7.9, pytest - 7.2.0, pluggy - 1.0.0
rootdir: E:\demo
plugins: assume - 2.4.3, rerunfailures - 10.2
collected 1 item

demo.py F                                                    [100%]

========================= FAILURES =========================
_____ [doctest] demo.demo.add _____
003
004       >>> add(10,20)
005       30
006       >>> add(100,200)
Expected:
    10
    :param a:
    :param b:
    :return:
Got:
    300

E:\demo\demo.py:6: DocTestFailure
================= short test summary info =================
FAILED demo.py::demo.demo.add
==================== 1 failed in 0.04s ====================

(demo - HCIhXOHq) E:\demo>
```

8.2 如何使用猴子补丁进行异常测试

有些场景下的测试可能需要修改全局配置或系统变量等操作,且这些操作仅仅是为了做一些测试,不希望永久修改,此时就需要使用猴子补丁了,猴子补丁,即 monkey-patch,是一个 fixture,它提供了以下方法。

```
monkeypatch.setattr(obj, name, value, raising = True)
monkeypatch.setattr("somemodule.obj.name", value, raising = True)
monkeypatch.delattr(obj, name, raising = True)
monkeypatch.setitem(mapping, name, value)
monkeypatch.delitem(obj, name, raising = True)
monkeypatch.setenv(name, value, prepend = None)
monkeypatch.delenv(name, raising = True)
```

```
monkeypatch.syspath_prepend(path)
monkeypatch.chdir(path)
```

我们可以通过猴子补丁修改 Path 的 home 属性,进而临时修改函数的功能,然后再进行测试,这样测试结束后,Path 的 home 属性并不会真的发生修改。

```
from pathlib import Path

def getssh():
    return Path.home() / ".ssh"
def test_getssh(monkeypatch):
    def mockreturn():
        return Path("/abc")
    monkeypatch.setattr(Path, "home", mockreturn)
    x = getssh()
    assert x == Path("/abc/.ssh")
def test_home():
    print(Path.home())
```

很明显,在 test_home 测试函数中,Path.home 属性并没有发生修改。

```
(demo-HCIhX0Hq) E:\demo> pytest -s
===================== test session starts =====================
platform win32 -- Python 3.7.9, pytest-7.2.0, pluggy-1.0.0
rootdir: E:\demo
plugins: assume-2.4.3, rerunfailures-10.2
collected 2 items

test_demo.py .C:\Users\Administrator
.

===================== 2 passed in 0.02s =====================

(demo-HCIhX0Hq) E:\demo>
```

假设 get_os_user_lower 函数为被测函数,用例中可以通过猴子补丁对变量进行临时设置或删除,这样就可以保证测试用例的准确性,否则当环境变量被修改或被删除后,用例的稳定性将会受到影响。

```
import os
import pytest


def get_os_user_lower():
    username = os.getenv("USER")
```

```python
    if username is None:
        raise OSError("USER environment is not set.")
    return username.lower()

def test_upper_to_lower(monkeypatch):
    monkeypatch.setenv("USER", "TestingUser")
    assert get_os_user_lower() == "testinguser"

def test_raise_exception(monkeypatch):
    monkeypatch.delenv("USER", raising=False)
    with pytest.raises(OSError):
        _ = get_os_user_lower()
```

我们看上述执行结果。

```
(demo-HCIhX0Hq) E:\demo> pytest -s
=================== test session starts ===================
platform win32 -- Python 3.7.9, pytest-7.2.0, pluggy-1.0.0
rootdir: E:\demo
plugins: assume-2.4.3, rerunfailures-10.2
collected 2 items

test_demo.py ..

=================== 2 passed in 0.02s ===================

(demo-HCIhX0Hq) E:\demo>
```

上述代码通过 fixture 继续优化，如下所示。

```python
import os
import pytest
def get_os_user_lower():
    username = os.getenv("USER")
    if username is None:
        raise OSError("USER environment is not set.")
    return username.lower()
@pytest.fixture
def mock_env_user(monkeypatch):
    monkeypatch.setenv("USER", "TestingUser")
@pytest.fixture
def mock_env_missing(monkeypatch):
    monkeypatch.delenv("USER", raising=False)
def test_upper_to_lower(mock_env_user):
    assert get_os_user_lower() == "testinguser"
```

```python
def test_raise_exception(mock_env_missing):
    with pytest.raises(OSError):
        _ = get_os_user_lower()
```

执行结果如下所示。修改仍然生效，而且代码更加整洁。

```
(demo-HCIhXOHq) E:\demo> pytest -s
==================== test session starts ====================
platform win32 -- Python 3.7.9, pytest-7.2.0, pluggy-1.0.0
rootdir: E:\demo
plugins: assume-2.4.3, rerunfailures-10.2
collected 2 items

test_demo.py ..

==================== 2 passed in 0.02s ====================

(demo-HCIhXOHq) E:\demo>
```

8.3 如何使测试过程中产生的文件自动删除

tmp_path 是一个 testcase 级别的 fixture，其返回的是 pathlib.Path 类型值，常用于创建一个独一无二的临时目录，主要用于测试写文件之类的场景，默认存放于系统的临时目录下，同时创建 pytest-N 目录，其中 N 是会不断自动加 1。

当我们测试创建文件，并往文件中写入测试用例时，倘若按照普通的写法，则必然存在一个问题，就是每执行一次就会产生一个文件，而这个文件是没有用的，但又不能每次都把创建的文件删除，因为如果这样，一旦测试脚本出现问题，就无法定位是哪里出了问题。这里使用 tmp_path 这个内置 fixture 就可以避免此问题的发生。

```python
def test_create_file(tmp_path):
    d = tmp_path / "sub"
    print(f"temp_dir:{d}")
    d.mkdir()
    p = d / "hello.txt"
    str_txt = "hello world"
    p.write_text(str_txt)
    assert p.read_text() == str_txt
    assert len(list(tmp_path.iterdir())) == 1
```

这时，我们可以根据打印的路径打开文件夹，观察 Pytest-x 文件夹的动态。如果是第一次使用此 fixture，则这里会显示 Pytest-0。

```
(demo-HCIhXOHq) E:\demo> pytest-s
==================== test session starts ====================
platform win32 -- Python 3.7.9, pytest-7.2.0, pluggy-1.0.0
rootdir: E:\demo
plugins: assume-2.4.3, rerunfailures-10.2
collected 1 item

test_demo.py
temp_dir:C:\Users\Administrator\AppData\Local\Temp\pytest-of-Administrator\pytest-0\test_create_file0\sub
.

==================== 1 passed in 0.04s ====================

(demo-HCIhXOHq) E:\demo>
```

这里,我们连续执行三次后,再观察文件夹,就会发现此时文件就剩 Pytest-2,Pytest-3,Pytest-4 三个了,Pytest-0 此时已经没有了,这就是 tmp_path 作用的结果,即自动保留最新的三个文件夹。这样就不怕系统会因为不断产生文件夹以及文件而崩溃的风险了。

tmp_path_factory 是一个 session 级别的 fixture,每次执行时,只会创建一个临时目录。

```
def test_create_file(tmp_path_factory):
    d = tmp_path_factory.mktemp("demo01") / "hello.txt"
    print(f"temp_dir:{d}")
    str_txt = "hello world"
    d.write_text(str_txt)
    assert d.read_text() == str_txt
def test_create_file2(tmp_path_factory):
    d = tmp_path_factory.mktemp("demo02") / "hello.txt"
    print(f"temp_dir:{d}")
    str_txt = "hello world"
    d.write_text(str_txt)
    assert d.read_text() == str_txt
```

虽然上面在两个用例中都调用了 fixture,但是其只创建了一个临时目录 Pytest-5,因为它是 session 级别的原因。

```
(demo-HCIhXOHq) E:\demo> pytest-s
==================== test session starts ====================
platform win32 -- Python 3.7.9, pytest-7.2.0, pluggy-1.0.0
rootdir: E:\demo
plugins: assume-2.4.3, rerunfailures-10.2
```

```
collected 2 items

test_demo.py
temp_dir:C:\Users\Administrator\AppData\Local\Temp\pytest-of-Administrator\pytest-5\demo010\hello.txt
.temp_dir:C:\Users\Administrator\AppData\Local\Temp\pytest-of-Administrator\pytest-5\demo020\hello.txt
.

==================== 2 passed in 0.03s ====================

(demo-HCIhXOHq) E:\demo>
```

tmpdir 和 tmp_path 功能是一样的，唯一区别是 tmpdir 返回的是 py.path.local 类型，而 tmp_path 返回的是 pathlib.Path 类型的，tmpdir 返回的值主要用于支持 os.path 的一些操作方法，另外，tmpdir 也是一个 testcase 级别的 fixture。

我们看如下的测试代码：

```python
def test_create_file(tmpdir):
    p = tmpdir.mkdir("sub").join("hello.txt")
    print(f"tmpdir:{p}")
    p.write("content")
    assert p.read() == "content"
    assert len(tmpdir.listdir()) == 1
```

执行结果如下所示。

```
(demo-HCIhXOHq) E:\demo> pytest -s
==================== test session starts ====================
platform win32 -- Python 3.7.9, pytest-7.2.0, pluggy-1.0.0
rootdir: E:\demo
plugins: assume-2.4.3, rerunfailures-10.2
collected 1 item

test_demo.py
tmpdir:C:\Users\Administrator\AppData\Local\Temp\pytest-of-Administrator\pytest-6\test_create_file0\sub\hello.txt
.

==================== 1 passed in 0.03s ====================

(demo-HCIhXOHq) E:\demo>
```

同样，tmpdir_factory 和 tmp_path_factory 功能也是如此，是一个 session 级别的

fixture,一次执行只会创建一个临时目录。

```
def test_create_file(tmpdir_factory):
    p = tmpdir_factory.mktemp("demo01").join("hello.txt")
    print(f"tmpdir:{p}")
    p.write("content")
    assert p.read() == "content"
def test_create_file2(tmpdir_factory):
    p = tmpdir_factory.mktemp("demo02").join("hello.txt")
    print(f"tmpdir:{p}")
    p.write("content")
    assert p.read() == "content"
```

可以发现,这里,同样执行了两个脚本,只创建了一个 Pytest-7 的目录,因为这个 fixture 是 session 级别的。

```
(demo-HCIhXOHq) E:\demo> pytest -s
==================== test session starts ====================
platform win32 -- Python 3.7.9, pytest-7.2.0, pluggy-1.0.0
rootdir: E:\demo
plugins: assume-2.4.3, rerunfailures-10.2
collected 2 items

test_demo.py
tmpdir:C:\Users\Administrator\AppData\Local\Temp\pytest-of-Administrator\pytest-7\demo010\hello.txt
.tmpdir:C:\Users\Administrator\AppData\Local\Temp\pytest-of-Administrator\pytest-7\demo020\hello.txt
.

==================== 2 passed in 0.03s ====================

(demo-HCIhXOHq) E:\demo>
```

8.4 如何动态获取 Pytest.ini 中的配置以及命令行参数

Pytestconfig 是 Pytest 中一个内置的 fixture,通过 Pytestconfig 我们可以动态获取 Pytest.ini 中的配置以及命令行中的参数值,就先来看下如何使用 Pytestconfig 动态获取 Pytest.ini 中的配置。

在 Pytest.ini 中设置下面日志格式等配置。

[pytest]

```
log_cli = True
log_cli_level = info
log_cli_format = %(asctime)s | %(levelname)s | %(filename)s:%(lineno)s | %(message)s
log_cli_date_format = %Y-%m-%d %H:%M:%S

log_level = info
log_format = %(asctime)s | %(levelname)s | %(filename)s:%(lineno)s | %(message)s
log_date_format = %Y-%m-%d %H:%M:%S
```

在测试用例 test_demo.py 的 test_01 中,调用 getini 方法获取 log_cli 以及 log_cli_level 的值,并将其打印出来,测试代码如下所示。

```
def test_01(pytestconfig):
    print(pytestconfig.getini("log_cli"))
    print(pytestconfig.getini("log_cli_level"))
```

可以看出,此时已经将 log_cli 的结果 True 及 log_cli_level 的结果 info 打印出来了。使用这个功能,我们可以在测试用例时针对一些特殊的配置做一些特殊处理,如此一来对测试用例的控制就更加容易了。

```
(demo-HCIhXOHq) E:\demo> pytest -s
True
info
.
--------------------------------------------------------------
Ran 1 tests in 0.02s

OK

(demo-HCIhXOHq) E:\demo>
```

此外,Pytestconfig 还可以动态获取命令行参数的值,比如我们给测试用例增加 mark 标签,然后在测试用例中尝试获取 -m 参数的对应值。

```
import pytest


@pytest.mark.smoke
def test_01(pytestconfig):
    print(pytestconfig.getoption("-m"))
```

我们通过 Pytest -s -m smoke 命令执行脚本,此时 -m 参数对应的值为 smoke,可以看到获取到了命令行的参数,并且打印出来了。因此,对于命令行的参数,在一些特殊场景下可以根据命令行做一些特殊处理,这样就大大增强了用例的灵活性。当然,在实际自动化脚本开发的过程中,这个功能基本是用不上的,或基本不需要使用的。像这些相对特殊的高级应用一般会在设计开发自动化测试框架的时候才会用到,这里只作简单介绍。

```
(demo-HCIhX0Hq) E:\demo> pytest -s -m smoke
smoke
.
------------------------------------------------------------
Ran 1 tests in 0.02s

OK

(demo-HCIhX0Hq) E:\demo>
```

8.5 如何在运行中动态获取用例的属性

request 是 Pytest 中一个功能非常强大的内置 fixture,request 可以动态获取用例的所有信息,如用例的名称、用例所属的模块,用例的参数,用例的文件路径等。下面就通过一个简单实例演示一下如何通过 request 实现对用例的透彻解析。

首先,在 conftest.py 中新建一个自动执行 session 级别的 fixture,for 循环的每个循环变量 item 其实就是一个测试用例,获取的测试用例就很容易执行所有属性了。

```python
import pytest

@pytest.fixture(scope="session",autouse=True)
def session_fixture(request):
    for item in request.node.items:
        print(item.name)
        print(item.module)
        print(item.path)
```

我们在 test_demo.py 中编写两个测试用例,代码如下所示。

```python
def test_01():
    print("in test_01...")
    assert 1 == 1


def test_02():
    print("in test_02...")
    assert 1 == 1
```

可以看到,这里将测试用例 test_01、test_02 名称,测试用例所属的模块,以及测试用例所在的文件路径都打印出来了。

```
(demo-HCIhXOHq) E:\demo> pytest -s
test_01
<module 'demo.test_demo' from 'E:\\demo\\test_demo.py'>
E:\demo\test_demo.py
test_02
<module 'demo.test_demo' from 'E:\\demo\\test_demo.py'>
E:\demo\test_demo.py
in test_01...
.in test_02...
.
----------------------------------------------------------
Ran 2 tests in 0.02s

OK

(demo-HCIhXOHq) E:\demo>
```

这个功能应该说是非常强大的,如我们希望针对某一类测试脚本做一些公共的处理,如果没有这个功能,则很大可能要去修改用例,而有了这个功能,我们就可以在 conftest.py 中写一个 fixture,然后获取所有的测试用例,再根据测试用例的某一个特点将要修改的用例识别出来,最后去做一些特殊处理,这样就非常简单了。

第 9 章　parameterize 参数化及数据驱动

9.1　测试函数使用 parametrize 进行参数化

在前面介绍 fixture 的时候，已经介绍过参数化可以实现数据驱动的用法，但使用 fixture 实现参数化有一点复杂，因为需要额外再定义一个 fixture。其实，在 Pytest 中，还有一种更加简单的方式来实现参数化，这就是内置的 parametrize，第一个参数指定形参名，第二个参数为列表，提供三个 name 值供测试使用。

```
import pytest

@pytest.mark.parametrize("name",["张三","李四","王五"])
def test_eval(name):
    print("name:",name)
    assert 1 == 1
```

可以发现，这里通过 parametrize 提供了三个数据，在执行结束后显示了有三个用例通过，即每个参数就是一个用例，这其实就是自动化测试中常说的数据驱动，对同一个测试函数，提供多个数据进行测试，每个数据即能作为一个用例。如果不采用这种参数化的方法，则需要将测试函数复制成 N 份，然后在每个几乎完全相同的测试函数中修改参数，这样会出现严重的代码冗余问题。

```
(demo-HCIhXOHq) E:\demo> pytest -s
==================== test session starts ====================
platform win32 -- Python 3.7.9, pytest-7.2.0, pluggy-1.0.0
rootdir: E:\demo
plugins: assume-2.4.3, rerunfailures-10.2
collected 3 items

test_demo.py name：张三
.name：李四
.name：王五

==================== 3 passed in 0.02s ====================
```

(demo-HCIhXOHq) E:\demo>

当测试函数有多个形参时,在参数化的时候,在指定参数列表的字符串中,要使用逗号进行间隔,而在参数列表中则使用列表的集合,下面的测试函数有两个形参,name 和 age。

```python
import pytest

@pytest.mark.parametrize("name,age", [["张三",20],["李四",18],["王五",24]])
def test_eval(name,age):
    print("name:",name)
    print("age:",age)
    assert 1 == 1
```

可以看出,这里显示有三个测试用例通过,因为参数化列表提供了三组数据,注意每一组数据都要放入 list 中。

```
(demo-HCIhXOHq) E:\demo> pytest -s
==================== test session starts ====================
platform win32 -- Python 3.7.9, pytest-7.2.0, pluggy-1.0.0
rootdir: E:\demo
plugins: assume-2.4.3, rerunfailures-10.2
collected 3 items

test_demo.py name:张三
age:20
.name:李四
age:18
.name:王五
age:24
.

==================== 3 passed in 0.02s ====================
```

(demo-HCIhXOHq) E:\demo>

当然,参数化列表中每组数据也可以放在元组中,在 Python 语言开发中,其实对于函数的参数,我们更加希望采用元组类型,因为元组类型是不允许修改的,这样可以避免很多不必要的麻烦。

```python
import pytest

@pytest.mark.parametrize("name,age", [("张三",20),("李四",18),("王五",24)])
def test_eval(name,age):
    print("name:",name)
```

```
    print("age:",age)
    assert 1 == 1
```

执行结果如下所示,可以看出,此时数值仍然是正确的。

```
(demo-HCIhXOHq) E:\demo> pytest -s
==================== test session starts ====================
platform win32 -- Python 3.7.9, pytest-7.2.0, pluggy-1.0.0
rootdir: E:\demo
plugins: assume-2.4.3, rerunfailures-10.2
collected 3 items

test_demo.py name:张三
age:20
.name:李四
age:18
.name:王五
age:24
.

==================== 3 passed in 0.02s ====================

(demo-HCIhXOHq) E:\demo>
```

9.2 测试类使用 parametrize 进行参数化

上一节介绍了测试函数的进行参数化的用法,本小节继续介绍参数化作用域测试类的使用方法。所谓作用于测试类,即作用于测试类中的所有方法,根据参数化在测试函数上的使用方式可知,测试类中的所有测试方法的形参必须是一致的,其具体使用方法如下所示。

```python
import pytest
@pytest.mark.parametrize("n,expected", [(1, 2),(3, 4),(1,10)])
class TestClass:
    def test_simple_case(self, n, expected):
        assert n + 1 == expected
    def test_weird_simple_case(self, n, expected):
        assert (n * 1) + 1 == expected
```

可以看到,当前测试类中有两个测试方法,参数化提供了三组数据,相当于对每个测试方法都运用了三组数据,总共有 6 个测试用例,由于有一组数据对这两个测试方法都是失败的,因此,执行结果显示有 4 个通过,2 个失败。

```
(demo-HCIhX0Hq) E:\demo> pytest -s
==================== test session starts ====================
platform win32 -- Python 3.7.9, pytest-7.2.0, pluggy-1.0.0
rootdir: E:\demo
plugins: assume-2.4.3, rerunfailures-10.2
collected 6 items

test_demo.py ..F..F

========================== FAILURES ==========================
_____ TestClass.test_simple_case[1-10] _____

self = <demo.test_demo.TestClass object at 0x000001F3353A7C08>
n = 1, expected = 10

    def test_simple_case(self, n, expected):
>       assert n + 1 == expected
E       assert (1 + 1) == 10

test_demo.py:5: AssertionError
_____ TestClass.test_weird_simple_case[1-10] _____

self = <demo.test_demo.TestClass object at 0x000001F33539FBC8>
n = 1, expected = 10

    def test_weird_simple_case(self, n, expected):
>       assert (n * 1) + 1 == expected
E       assert ((1 * 1) + 1) == 10

test_demo.py:7: AssertionError
================== short test summary info ==================
FAILED test_demo.py::TestClass::test_simple_case[1-10] - assert (1 + 1) == 10
FAILED test_demo.py::TestClass::test_weird_simple_case[1-10] - assert ((1 * 1) + 1) == 10
================ 2 failed, 4 passed in 0.08s ================

(demo-HCIhX0Hq) E:\demo>
```

9.3 通过 pytestmark 对测试模块内的代码进行参数化

通过在模块内定义 Pytestmark 变量，可以对模块内的所有测试函数以及测试类中的测试方法进行参数化处理，如定义了一个 Pytestmark 变量，然后又定义了一个测试函数，同时定义了一个测试类，类中又定义了两个测试方法，如果要在测试文件中进行参数化，则必须要求测试函数以及测试类中测试方法的形参应是一致的。

```python
import pytest

pytestmark = pytest.mark.parametrize("n,expected", [(1, 2),(3, 4),(1,100)])

def test_01(n,expected):
    assert n + 1 == expected

class TestClass:
    def test_02(self, n, expected):
        assert n + 1 == expected

    def test_03(self, n, expected):
        assert (n * 1) + 1 == expected
```

可见，此时测试文件中的测试函数和测试方法均执行了，且对参数化提供的3组数据进行了验证，从结果看，有6个用例通过，有3个用例失败。

```
(demo-HCIhX0Hq) E:\demo> pytest -s
==================== test session starts ====================
platform win32 -- Python 3.7.9, pytest-7.2.0, pluggy-1.0.0
rootdir: E:\demo
plugins: assume-2.4.3, rerunfailures-10.2
collected 9 items

test_demo.py ..F..F..F

========================= FAILURES =========================
_____ test_01[1-100] _____

n = 1, expected = 100

    def test_01(n,expected):
>       assert n + 1 == expected
E       assert (1 + 1) == 100

test_demo.py:6: AssertionError
_____ TestClass.test_02[1-100] _____

self = <demo.test_demo.TestClass object at 0x0000022C5E04D408>
n = 1, expected = 100
```

```
        def test_02(self, n, expected):
>           assert n + 1 == expected
E           assert (1 + 1) == 100

test_demo.py:10: AssertionError
_____ TestClass.test_03[1 - 100] _____

self = <demo.test_demo.TestClass object at 0x0000022C5E04DFC8>
n = 1, expected = 100

        def test_03(self, n, expected):
>           assert (n * 1) + 1 == expected
E           assert ((1 * 1) + 1) == 100

test_demo.py:13: AssertionError
================= short test summary info =================
FAILED test_demo.py::test_01[1 - 100] - assert (1 + 1) == 100
FAILED test_demo.py::TestClass::test_02[1 - 100] - assert (1 + 1) == 100
FAILED test_demo.py::TestClass::test_03[1 - 100] - assert ((1 * 1) + 1) == 100
================= 3 failed, 6 passed in 0.09s =================

(demo - HCIhX0Hq) E:\demo>
```

9.4　parametrize 参数化时使用 skip 标记

在参数化过程中,我们也可以使用 skip 对特定的数据进行标记,如由于某种原因,当前实现的功能尚未满足(1,100)这组数据,此时就可以使用 skip 进行标记,标记方式如下所示。

```
import pytest

pytestmark = pytest.mark.parametrize("n,expected", [(1, 2), (3, 4), pytest.param(1, 100, marks = pytest.mark.skip)])

def test_01(n, expected):
    assert n + 1 == expected


class TestClass:
    def test_02(self, n, expected):
        assert n + 1 == expected
```

```
    def test_03(self, n, expected):
        assert (n * 1) + 1 == expected
```

可以看出,此时有 3 个用例标记了 skipped,这是因为(1,10)这组数据被 skip 标记,而这组数据是作用于当前模块中的测试函数中的。

```
(demo-HCIhXOHq) E:\demo> pytest -s
==================== test session starts ====================
platform win32 -- Python 3.7.9, pytest-7.2.0, pluggy-1.0.0
rootdir: E:\demo
plugins: assume-2.4.3, rerunfailures-10.2
collected 9 items

test_demo.py ..s..s..s

============== 6 passed, 3 skipped in 0.03s ===============

(demo-HCIhXOHq) E:\demo>
```

9.5 parametrize 参数化时使用 xfail 标记

在 parametrize 参数化过程中同样可以使用 xfail 标记测试用例,如将(1,100)这组数据标记为 xfail。

```python
import pytest

pytestmark = pytest.mark.parametrize("n,expected", [(1, 2), (3, 4), pytest.param(1, 100, marks=pytest.mark.xfail)])

def test_01(n, expected):
    assert n + 1 == expected

class TestClass:
    def test_02(self, n, expected):
        assert n + 1 == expected

    def test_03(self, n, expected):
        assert (n * 1) + 1 == expected
```

从执行结果可以看出,此时有 3 个测试用例被标记为 xfail,有 6 个用例通过。

```
(demo-HCIhXOHq) E:\demo> pytest -s
==================== test session starts ====================
```

```
platform win32 -- Python 3.7.9, pytest-7.2.0, pluggy-1.0.0
rootdir: E:\demo
plugins: assume-2.4.3, rerunfailures-10.2
collected 9 items

test_demo.py ..x..x..x

=============== 6 passed, 3 xfailed in 0.08s ================

(demo-HCIhXOHq) E:\demo>
```

9.6 parametrize 参数化时对两组数据进行全排列组合测试

Parametrize 参数化同样支持对两组数据全排列组合测试,这个功能对一些要求数据进行全排列覆盖测试的场景是非常有用的,具体用法如下所示,使用两次参数化声明即可完成。

```python
import pytest

@pytest.mark.parametrize("x", [0, 1])
@pytest.mark.parametrize("y", [2, 3])
def test_foo(x, y):
    print("x:",x)
    print("y:",y)
    assert x < y
```

这里可以看出,x 和 y 分别进行了全排列组合,即(0,2),(0,3),(1,2),(1,3),执行结果也明确显示有 4 个测试用例。

```
(demo-HCIhXOHq) E:\demo> pytest -s
==================== test session starts ====================
platform win32 -- Python 3.7.9, pytest-7.2.0, pluggy-1.0.0
rootdir: E:\demo
plugins: assume-2.4.3, rerunfailures-10.2
collected 4 items

test_demo.py x: 0
y: 2
.x: 1
y: 2
.x: 0
y: 3
```

```
.x: 1
y: 3
.
================== 4 passed in 0.02s ==================

(demo-HCIhXOHq) E:\demo>
```

其中一组数据是单个类型,另一组数据是多个类型,我们看两个类型的情况,再将这两组数据放在一起进行全排列组合,具体使用方法见如下代码。

```python
import pytest

@pytest.mark.parametrize("x", [0, 1])
@pytest.mark.parametrize("y,z", [(2,3), (3,4)])
def test_foo(x, y, z):
    print("x:",x)
    print("y:",y)
    print("z:",z)
    assert x + y == z
```

执行结果如下所示,其使用方式基本是类似的。

```
(demo-HCIhXOHq) E:\demo> pytest -s
================== test session starts ==================
platform win32 -- Python 3.7.9, pytest-7.2.0, pluggy-1.0.0
rootdir: E:\demo
plugins: assume-2.4.3, rerunfailures-10.2
collected 4 items

test_demo.py x: 0
y: 2
z: 3
Fx: 1
y: 2
z: 3
.x: 0
y: 3
z: 4
Fx: 1
y: 3
z: 4
.
```

```
=========================== FAILURES ===========================
_____ test_foo[2-3-0] _____

x = 0, y = 2, z = 3

    @pytest.mark.parametrize("x", [0, 1])
    @pytest.mark.parametrize("y,z", [(2,3), (3,4)])
    def test_foo(x, y,z):
        print("x:",x)
        print("y:",y)
        print("z:",z)
>       assert x + y == z
E       assert (0 + 2) == 3

test_demo.py:9: AssertionError
_____ test_foo[3-4-0] _____

x = 0, y = 3, z = 4

    @pytest.mark.parametrize("x", [0, 1])
    @pytest.mark.parametrize("y,z", [(2,3), (3,4)])
    def test_foo(x, y,z):
        print("x:",x)
        print("y:",y)
        print("z:",z)
>       assert x + y == z
E       assert (0 + 3) == 4

test_demo.py:9: AssertionError
==================== short test summary info ====================
FAILED test_demo.py::test_foo[2-3-0] - assert (0 + 2) == 3
FAILED test_demo.py::test_foo[3-4-0] - assert (0 + 3) == 4
================= 2 failed, 2 passed in 0.08s =================

(demo-HCIhX0Hq) E:\demo>
```

同理,假如有三组数据需要进行全排列组合,只需要再增加一个个参数化的声明即可。

```
import pytest

@pytest.mark.parametrize("x", [0, 1])
@pytest.mark.parametrize("y", [2, 3])
```

```
@pytest.mark.parametrize("z",[4,5])
def test_foo(x,y,z):
    print("x:",x)
    print("y:",y)
    print("z:",z)
    assert x + y == z
```

可以发现,此时的组合个数为 2×2×2=8 个,总共有 8 个用例。

```
(demo-HCIhX0Hq) E:\demo> pytest -s
==================== test session starts ====================
platform win32 -- Python 3.7.9, pytest-7.2.0, pluggy-1.0.0
rootdir: E:\demo
plugins: assume-2.4.3, rerunfailures-10.2
collected 8 items

test_demo.py x: 0
y: 2
z: 4
Fx: 1
y: 2
z: 4
Fx: 0
y: 3
z: 4
Fx: 1
y: 3
z: 4
.x: 0
y: 2
z: 5
Fx: 1
y: 2
z: 5
Fx: 0
y: 3
z: 5
Fx: 1
y: 3
z: 5
F


========================= FAILURES =========================
_____ test_foo[4-2-0] _____
```

```
x = 0, y = 2, z = 4

        @pytest.mark.parametrize("x", [0, 1])
        @pytest.mark.parametrize("y", [2, 3])
        @pytest.mark.parametrize("z", [4, 5])
        def test_foo(x, y, z):
            print("x:",x)
            print("y:",y)
            print("z:",z)
>           assert x + y == z
E           assert (0 + 2) == 4

test_demo.py:10: AssertionError
_____ test_foo[4-2-1] _____

x = 1, y = 2, z = 4

        @pytest.mark.parametrize("x", [0, 1])
        @pytest.mark.parametrize("y", [2, 3])
        @pytest.mark.parametrize("z", [4, 5])
        def test_foo(x, y, z):
            print("x:",x)
            print("y:",y)
            print("z:",z)
>           assert x + y == z
E           assert (1 + 2) == 4

test_demo.py:10: AssertionError
_____ test_foo[4-3-0] _____

x = 0, y = 3, z = 4

        @pytest.mark.parametrize("x", [0, 1])
        @pytest.mark.parametrize("y", [2, 3])
        @pytest.mark.parametrize("z", [4, 5])
        def test_foo(x, y, z):
            print("x:",x)
            print("y:",y)
            print("z:",z)
>           assert x + y == z
E           assert (0 + 3) == 4
```

```
test_demo.py:10: AssertionError
_____ test_foo[5-2-0] _____
```

x = 0, y = 2, z = 5

```
    @pytest.mark.parametrize("x", [0, 1])
    @pytest.mark.parametrize("y", [2, 3])
    @pytest.mark.parametrize("z", [4, 5])
    def test_foo(x, y,z):
        print("x:",x)
        print("y:",y)
        print("z:",z)
>       assert x + y == z
E       assert (0 + 2) == 5
```

```
test_demo.py:10: AssertionError
_____ test_foo[5-2-1] _____
```

x = 1, y = 2, z = 5

```
    @pytest.mark.parametrize("x", [0, 1])
    @pytest.mark.parametrize("y", [2, 3])
    @pytest.mark.parametrize("z", [4, 5])
    def test_foo(x, y,z):
        print("x:",x)
        print("y:",y)
        print("z:",z)
>       assert x + y == z
E       assert (1 + 2) == 5
```

```
test_demo.py:10: AssertionError
_____ test_foo[5-3-0] _____
```

x = 0, y = 3, z = 5

```
    @pytest.mark.parametrize("x", [0, 1])
    @pytest.mark.parametrize("y", [2, 3])
    @pytest.mark.parametrize("z", [4, 5])
    def test_foo(x, y,z):
        print("x:",x)
        print("y:",y)
        print("z:",z)
>       assert x + y == z
```

```
E       assert (0 + 3) == 5

test_demo.py:10: AssertionError
_____ test_foo[5-3-1] _____
x = 1, y = 3, z = 5

    @pytest.mark.parametrize("x", [0, 1])
    @pytest.mark.parametrize("y", [2, 3])
    @pytest.mark.parametrize("z", [4, 5])
    def test_foo(x, y, z):
        print("x:", x)
        print("y:", y)
        print("z:", z)
>       assert x + y == z
E       assert (1 + 3) == 5

test_demo.py:10: AssertionError
================= short test summary info =================
FAILED test_demo.py::test_foo[4-2-0] - assert (0 + 2) == 4
FAILED test_demo.py::test_foo[4-2-1] - assert (1 + 2) == 4
FAILED test_demo.py::test_foo[4-3-0] - assert (0 + 3) == 4
FAILED test_demo.py::test_foo[5-2-0] - assert (0 + 2) == 5
FAILED test_demo.py::test_foo[5-2-1] - assert (1 + 2) == 5
FAILED test_demo.py::test_foo[5-3-0] - assert (0 + 3) == 5
FAILED test_demo.py::test_foo[5-3-1] - assert (1 + 3) == 5
================ 7 failed, 1 passed in 0.13s ================

(demo-HCIhXOHq) E:\demo>
```

第 10 章 告 警

10.1 如何使用命令行控制告警

在执行自动化脚本的时候,出现告警的情况非常常见,比如我们使用了一个很快要被废弃了的语法,又如我们用了一个不被推荐的用法等,我们可以对告警不做任何处理,但我们有必要了解一下有哪些方法处理告警,如下面的代码,这里使用了一个未声明的 mark 标签来进行处理。

```
import pytest

@pytest.mark.smoke
def test_demo1():
    print("in test_demo1 ...")
    assert 1 == 1
```

我们看到,这里就产生了一个告警,提示 smoke 这个标签未注册,告警中还给出了参考文档的链接。

```
(demo-HCIhX0Hq) E:\demo> pytest -s
==================== test session starts ====================
platform win32 -- Python 3.7.9, pytest-7.2.0, pluggy-1.0.0
rootdir: E:\demo
plugins: assume-2.4.3, rerunfailures-10.2
collected 1 item

test_demo.py in test_demo1 ...
.

==================== warnings summary ====================
test_demo.py:3
  E:\demo\test_demo.py:3: PytestUnknownMarkWarning: Unknown pytest.mark.smoke - is this a typo? You can register custom marks to avoid this warning - for details, see https://docs.pytest.org/en/stable/how-to/mark.html
    @pytest.mark.smoke
```

```
--Docs: https://docs.pytest.org/en/stable/how-to/capture-warnings.html
=============== 1 passed, 1 warning in 0.02s ===============

(demo-HCIhX0Hq) E:\demo>
```

从执行的结果可以看出，告警信息较多，不利于问题定位以及回显显示，所以在有些情况下，我们可能希望不显示这些告警信息，那么使用-W ignore 可以做到，如下所示。

```
(demo-HCIhX0Hq) E:\demo> pytest -W ignore
=================== test session starts ===================
platform win32 -- Python 3.7.9, pytest-7.2.0, pluggy-1.0.0
rootdir: E:\demo
plugins: assume-2.4.3, rerunfailures-10.2
collected 1 item

test_demo.py .                                       [100%]

=================== 1 passed in 0.02s ===================

(demo-HCIhX0Hq) E:\demo>
```

当然，从另外一个角度来看，有告警说明最好做适当的修改或调整，可以使用-W error，将所有的告警信息换为报错，即强制使测试用例失败，因为只有用例失败了，脚本开发者才会作出对应的修改和调整。

这里就将测试 smoke 的标签显示为错误，这样一来，脚本开发者就必须去解决所警告的问题。当然是否要采用这种策略，要根据具体的情况而定。

```
(demo-HCIhX0Hq) E:\demo> pytest -W error
=================== test session starts ===================
platform win32 -- Python 3.7.9, pytest-7.2.0, pluggy-1.0.0
rootdir: E:\demo
plugins: assume-2.4.3, rerunfailures-10.2
collected 0 items / 1 error

========================= ERRORS =========================
_____ ERROR collecting test_demo.py _____
test_demo.py:3: in <module>
    @pytest.mark.smoke
C:\Users\Administrator\.virtualenvs\demo-HCIhX0Hq\lib\site-packages\_pytest\mark\structures.py:549: in __getattr__
    2,
E   pytest.PytestUnknownMarkWarning: Unknown pytest.mark.smoke - is this a typo?  You can register custom marks to avoid this warning - for details, see https://docs.pytest.org/en/
```

```
stable/how-to/mark.html
=================== short test summary info ===================
ERROR test_demo.py - pytest.PytestUnknownMarkWarning: Unknown pytest.mark.sm...
!!!!!!!!! Interrupted: 1 error during collection !!!!!!!!!
===================== 1 error in 0.11s =====================

(demo-HCIhXOHq) E:\demo>
```

上面的对告警处理方法是将所有的告警进行统一处理,即当使用-W ignore时,所有类型的告警都将不显示,当使用-W error时,所有类型的告警都将报错。在实际应用中,我们可能会遇到这种情况,即希望不显示一些特定类型的告警,或者说希望特定类型的告警报错,此时就需要在-W ignore 和-W error 后面指定告警类型。

在如下测试代码中,一种是 smoke 标签这种未注册类型的告警,一种是人为抛出的用户类型告警。

```python
import pytest
import warnings


@pytest.mark.smoke
def test_demo1():
    print("in test_demo1 ...")
    warnings.warn(SyntaxWarning("warning,used to test..."))
    assert 1 == 1
```

这里显示有两个告警,从执行结果中可以看到,一个告警类型是 PytestUnknownMarkWarning,另一个告警类型是 SyntaxWarning。

```
(demo-HCIhXOHq) E:\demo> pytest -s
===================== test session starts =====================
platform win32 -- Python 3.7.9, pytest-7.2.0, pluggy-1.0.0
rootdir: E:\demo
plugins: assume-2.4.3, rerunfailures-10.2
collected 1 item

test_demo.py in test_demo1 ...
.

===================== warnings summary =====================
test_demo.py:4
  E:\demo\test_demo.py:4: PytestUnknownMarkWarning: Unknown pytest.mark.smoke - is this a typo?  You can register custom marks to avoid this warning - for details, see https://docs.pytest.org/en/stable/how-to/mark.html
    @pytest.mark.smoke
```

```
test_demo.py::test_demo1
  E:\demo\test_demo.py:7: SyntaxWarning: warning,used to test...
    warnings.warn(SyntaxWarning("warning,used to test..."))

-- Docs: https://docs.pytest.org/en/stable/how-to/capture-warnings.html
=============== 1 passed, 2 warnings in 0.02s ===============

(demo-HCIhXOHq) E:\demo>
```

比如,这里不想显示 PytestUnknownMarkWarning,但希望 SyntaxWarning 告警显示正常,此时就需要指定告警类型精确不显示了。这是 Python 内置的告警类型问题,在 Python 语言中,Warning 是所有告警的父类,如果指定 Warning 类型,则所有告警将不显示,如下所示。

```
(demo-HCIhXOHq) E:\demo> pytest -W ignore::Warning
==================== test session starts ====================
platform win32 -- Python 3.7.9, pytest-7.2.0, pluggy-1.0.0
rootdir: E:\demo
plugins: assume-2.4.3, rerunfailures-10.2
collected 1 item

test_demo.py .                                        [100%]

==================== 1 passed in 0.02s ====================

(demo-HCIhXOHq) E:\demo>
```

子类告警类型主要有 BytesWarning、DeprecationWarning、FutureWarning、ImportWarning、PendingDeprecationWarning、ResourceWarning、RuntimeWarning、SyntaxWarning、UnicodeWarning、UserWarning,显然 Python 内置告警类型中没有 PytestUnknownMarkWarning,那么这里如果直接指定 PytestUnknownMarkWarning 类型是会报错的,如下所示。

```
(demo-HCIhXOHq) E:\demo> pytest -W ignore::PytestUnknownMarkWarning
ERROR: while parsing the following warning configuration:

  ignore::PytestUnknownMarkWarning

This error occurred:

Traceback (most recent call last):
  File "C:\Users\Administrator\.virtualenvs\demo-HCIhXOHq\lib\site-packages\_pytest\config\__init__.py", line 1690, in parse_warning_filter
```

```
        category: Type[Warning] = _resolve_warning_category(category_)
    File "C:\Users\Administrator\.virtualenvs\demo-HCIhXOHq\lib\site-packages\_pytest
\config\__init__.py", line 1729, in _resolve_warning_category
        cat = getattr(m, klass)
AttributeError: module 'builtins' has no attribute 'PytestUnknownMarkWarning'


(demo-HCIhXOHq) E:\demo>
```

实质上 PytestUnknownMarkWarning 是 UserWarning 的子类,因此这里只需要指定 UserWarning 类型即可,即此时执行结果只显示脚本中主动抛出的 SyntaxWarning 的告警了。

```
(demo-HCIhXOHq) E:\demo> pytest -W ignore::UserWarning
==================== test session starts ====================
platform win32 -- ython 3.7.9, pytest-7.2.0, pluggy-1.0.0
rootdir: E:\demo
plugins: assume-2.4.3, rerunfailures-10.2
collected 1 item

test_demo.py .                                        [100%]

==================== warnings summary ====================
test_demo.py::test_demo1
  E:\demo\test_demo.py:7: SyntaxWarning: warning,used to test...
    warnings.warn(SyntaxWarning("warning,used to test..."))

-- Docs: https://docs.pytest.org/en/stable/how-to/capture-warnings.html
============== 1 passed, 1 warning in 0.02s ================


(demo-HCIhXOHq) E:\demo>
```

同理,如果希望脚本中主动抛出的 SyntaxWarning 类型的告警为错误,同样在 -W error 后指定即可,如下所示。

```
(demo-HCIhXOHq) E:\demo> pytest -W error::SyntaxWarning
==================== test session starts ====================
platform win32 -- Python 3.7.9, pytest-7.2.0, pluggy-1.0.0
rootdir: E:\demo
plugins: assume-2.4.3, rerunfailures-10.2
collected 1 item

test_demo.py F                                        [100%]
```

```
========================= FAILURES =========================
_____ test_demo1 _____

        @pytest.mark.smoke
        def test_demo1():
            print("in test_demo1 ...")
>           warnings.warn(SyntaxWarning("warning,used to test..."))
E           SyntaxWarning: warning,used to test...

test_demo.py:7: SyntaxWarning
------------------- Captured stdout call -------------------
in test_demo1 ...
==================== warnings summary ====================
test_demo.py:4
  E:\demo\test_demo.py:4: PytestUnknownMarkWarning: Unknown pytest.mark.smoke - is this a typo?  You can register custom marks to avoid this warning - for details, see https://docs.pytest.org/en/stable/how-to/mark.html
    @pytest.mark.smoke

-- Docs: https://docs.pytest.org/en/stable/how-to/capture-warnings.html
================= short test summary info =================
FAILED test_demo.py::test_demo1 - SyntaxWarning: warning,used to test...
=============== 1 failed, 1 warning in 0.07s ===============

(demo-HCIhXOHq) E:\demo>
```

10.2　如何通过 filterwarnings 配置告警或将告警报错

首先,我们在脚本中人工抛出一条告警,测试代码如下所示。

```
import pytest
import warnings

def test_demo1():
    print("in test_demo1 ...")
    warnings.warn(SyntaxWarning("warning,used to test..."))
    assert 1 == 1
```

执行结果如下所示,即这里在执行结果中也显示了一条告警信息。

```
(demo-HCIhXOHq) E:\demo> pytest -s
```

```
=================== test session starts ===================
platform win32 -- Python 3.7.9, pytest-7.2.0, pluggy-1.0.0
rootdir: E:\demo
plugins: assume-2.4.3, rerunfailures-10.2
collected 1 item


test_demo.py in test_demo1 ...
.


=================== warnings summary ===================
test_demo.py::test_demo1
  E:\demo\test_demo.py:7: SyntaxWarning: warning,used to test...
    warnings.warn(SyntaxWarning("warning,used to test..."))

-- Docs: https://docs.pytest.org/en/stable/how-to/capture-warnings.html
=============== 1 passed, 1 warning in 0.02s ===============

(demo-HCIhXOHq) E:\demo>
```

在脚本中,我们可以通过filterwarnings配置告警信息的处理方式,比如将其忽略不显示,即此时在测试函数中加一个"装饰器装饰",在装饰器中使用ignore,注意当按照告警类型忽略不显示的时候,ignore后面是两个冒号。

```
import pytest
import warnings

@pytest.mark.filterwarnings("ignore::SyntaxWarning")
def test_demo1():
    print("in test_demo1 ...")
    warnings.warn(SyntaxWarning("warning,used to test..."))
    assert 1 == 1
```

执行结果如下所示,即不再有告警信息。

```
(demo-HCIhXOHq) E:\demo> pytest -s
=================== test session starts ===================
platform win32 -- Python 3.7.9, pytest-7.2.0, pluggy-1.0.0
rootdir: E:\demo
plugins: assume-2.4.3, rerunfailures-10.2
collected 1 item


test_demo.py in test_demo1 ...
.
```

```
=================== 1 passed in 0.02s ===================

(demo-HCIhXOHq) E:\demo>
```

同理,若将告警转换为报错,只需要将代码中装饰器中的 ignore 修改为 error 即可,如下所示。

```python
import pytest
import warnings


@pytest.mark.filterwarnings("error::SyntaxWarning")
def test_demo1():
    print("in test_demo1 ...")
    warnings.warn(SyntaxWarning("warning,used to test..."))
    assert 1 == 1
```

执行结果如下,可以看到,此时用例失败了,原因就是这个告警的问题。

```
(demo-HCIhXOHq) E:\demo> pytest -s
=================== test session starts ===================
platform win32 -- Python 3.7.9, pytest-7.2.0, pluggy-1.0.0
rootdir: E:\demo
plugins: assume-2.4.3, rerunfailures-10.2
collected 1 item

test_demo.py in test_demo1 ...
F


========================= FAILURES =========================
_____ test_demo1 _____

    @pytest.mark.filterwarnings("error::SyntaxWarning")
    def test_demo1():
        print("in test_demo1 ...")
>       warnings.warn(SyntaxWarning("warning,used to test..."))
E       SyntaxWarning: warning,used to test...

test_demo.py:7: SyntaxWarning
================= short test summary info =================
FAILED test_demo.py::test_demo1 - SyntaxWarning: warning,used to test...
=================== 1 failed in 0.07s ===================

(demo-HCIhXOHq) E:\demo>
```

除上述对告警配置以外,还可以通过对告警信息进行正则匹配,如使用 ignore 时,

ignore 后面只跟一个冒号，冒号后面的内容就是对告警信息的正则表达式。如 test_demo1 中匹配告警信息中有 test 字符的就忽略，而 test_demo2 中匹配告警信息中有 demo 字符的忽略。从下面代码分析，可知 test_demo2 中的告警信息无法匹配，即按照理论分析，test_demo1 中的告警会忽略不显示，而 test_demo2 中的告警会显示。

```
import pytest
import warnings

@pytest.mark.filterwarnings("ignore:.*test.*")
def test_demo1():
    print("in test_demo1 ...")
    warnings.warn(SyntaxWarning("warning,used to test..."))
    assert 1 == 1

@pytest.mark.filterwarnings("ignore:.*demo.*")
def test_demo2():
    print("in test_demo2 ...")
    warnings.warn(SyntaxWarning("warning,used to test..."))
    assert 1 == 1
```

可以看出，执行结果确实如此，test_demo2 中的告警之后报出来了，而 test_demo1 中的告警并未显示。

```
(demo-HCIhXOHq) E:\demo> pytest -s
==================== test session starts ====================
platform win32 -- Python 3.7.9, pytest-7.2.0, pluggy-1.0.0
rootdir: E:\demo
plugins: assume-2.4.3, rerunfailures-10.2
collected 2 items

test_demo.py in test_demo1 ...
. in test_demo2 ...
.


==================== warnings summary ====================
test_demo.py::test_demo2
  E:\demo\test_demo.py:13: SyntaxWarning: warning,used to test...
    warnings.warn(SyntaxWarning("warning,used to test..."))

-- Docs: https://docs.pytest.org/en/stable/how-to/capture-warnings.html
=============== 2 passed, 1 warning in 0.02s ===============

(demo-HCIhXOHq) E:\demo>
```

同理，将告警转换为报错信息，是同样支持正则表达式的，比如如下代码。

```python
import pytest
import warnings


@pytest.mark.filterwarnings("error:.*test.*")
def test_demo1():
    print("in test_demo1...")
    warnings.warn(SyntaxWarning("warning,used to test..."))
    assert 1 == 1


@pytest.mark.filterwarnings("error:.*demo.*")
def test_demo2():
    print("in test_demo2...")
    warnings.warn(SyntaxWarning("warning,used to test..."))
    assert 1 == 1
```

从执行的结果看出，test_demo1 中的告警与正则表达匹配，因此报错了，而 test_demo2 中的告警并未匹配到正则表达式，因此不会报错，所以之后报出了告警信息。

```
(demo-HCIhX0Hq) E:\demo> pytest -s
==================== test session starts ====================
platform win32 -- Python 3.7.9, pytest-7.2.0, pluggy-1.0.0
rootdir: E:\demo
plugins: assume-2.4.3, rerunfailures-10.2
collected 2 items

test_demo.py in test_demo1...
Fin test_demo2...
.

========================== FAILURES ==========================
_____ test_demo1 _____

    @pytest.mark.filterwarnings("error:.*test.*")
    def test_demo1():
        print("in test_demo1...")
>       warnings.warn(SyntaxWarning("warning,used to test..."))
E       SyntaxWarning: warning,used to test...

test_demo.py:7: SyntaxWarning
==================== warnings summary ====================
test_demo.py::test_demo2
  E:\demo\test_demo.py:13: SyntaxWarning: warning,used to test...
```

```
        warnings.warn(SyntaxWarning("warning,used to test..."))
```

--Docs：https://docs.pytest.org/en/stable/how-to/capture-warnings.html
=================== short test summary info ==================
FAILED test_demo.py::test_demo1 - SyntaxWarning: warning,used to test...
========= 1 failed, 1 passed, 1 warning in 0.07s ==========

(demo-HCIhXOHq) E:\demo>

10.3 如何将一个测试文件产生的告警都忽略或者转换为报错

在有些场景下，我们需要对一个测试文件中所有测试方法中的告警忽略不显示或将所有的告警转换为错误，此时只需要在文件开头定义 Pytestmark 即可实现，如下代码中就是在定义 Pytestmark 中使用了 ignore 参数，表示当前文件中的所有告警都忽略，不显示。

```python
import pytest
import warnings
pytestmark = pytest.mark.filterwarnings("ignore")

def test_demo1():
    print("in test_demo1 ...")
    warnings.warn(SyntaxWarning("warning,used to test..."))
    assert 1 == 1


def test_demo2():
    print("in test_demo2 ...")
    warnings.warn(UserWarning("warning,used to demo..."))
    assert 1 == 1
```

可以看出此时文件中定义的两个告警均未显示。

```
(demo-HCIhXOHq) E:\demo> pytest -s
==================== test session starts ===================
platform win32 -- Python 3.7.9, pytest-7.2.0, pluggy-1.0.0
rootdir: E:\demo
plugins: assume-2.4.3, rerunfailures-10.2
collected 2 items

test_demo.py in test_demo1 ...
.in test_demo2 ...
```

```
==================== 2 passed in 0.02s ====================

(demo-HCIhXOHq) E:\demo>
```

如下所示,即定义 Pytestmark 的时候指定 error,则可将当前文件中的所有告警均转换为报错。

```
import pytest
import warnings
pytestmark = pytest.mark.filterwarnings("error")


def test_demo1():
    print("in test_demo1 ...")
    warnings.warn(SyntaxWarning("warning,used to test..."))
    assert 1 == 1


def test_demo2():
    print("in test_demo2 ...")
    warnings.warn(UserWarning("warning,used to demo..."))
    assert 1 == 1
```

这里两个用例均报错了,报错原因是因为告警。

```
(demo-HCIhXOHq) E:\demo> pytest -s
==================== test session starts ====================
platform win32 -- Python 3.7.9, pytest-7.2.0, pluggy-1.0.0
rootdir: E:\demo
plugins: assume-2.4.3, rerunfailures-10.2
collected 2 items

test_demo.py in test_demo1 ...
Fin test_demo2 ...
F

========================= FAILURES =========================
_____ test_demo1 _____

    def test_demo1():
        print("in test_demo1 ...")
>       warnings.warn(SyntaxWarning("warning,used to test..."))
E       SyntaxWarning: warning,used to test...
```

```
test_demo.py:7: SyntaxWarning
_____ test_demo2 _____

    def test_demo2():
        print("in test_demo2 ...")
>       warnings.warn(UserWarning("warning,used to demo..."))
E       UserWarning: warning,used to demo...

test_demo.py:13: UserWarning
================= short test summary info =================
FAILED test_demo.py::test_demo1 - SyntaxWarning: warning,used to test...
FAILED test_demo.py::test_demo2 - UserWarning: warning,used to demo...
==================== 2 failed in 0.08s ====================

(demo-HCIhXOHq) E:\demo>
```

同样，我们也可以通过指定告警类型对当前文件中的所有告警进行过滤，如下所示是对当前文件中的 UserWarning 类型的告警做了忽略。

```python
import pytest
import warnings

pytestmark = pytest.mark.filterwarnings("ignore::UserWarning")


def test_demo1():
    print("in test_demo1 ...")
    warnings.warn(SyntaxWarning("warning,used to test..."))
    assert 1 == 1


def test_demo2():
    print("in test_demo2 ...")
    warnings.warn(UserWarning("warning,used to demo..."))
    assert 1 == 1
```

可以看出，此时 UserWarning 的告警已经不显示了，而 SyntaxWarning 类型的告警依然是显示的。

```
(demo-HCIhXOHq) E:\demo> pytest -s
==================== test session starts ====================
platform win32 -- Python 3.7.9, pytest-7.2.0, pluggy-1.0.0
rootdir: E:\demo
plugins: assume-2.4.3, rerunfailures-10.2
collected 2 items
```

```
test_demo.py in test_demo1 ...
. in test_demo2 ...
.


=================== warnings summary ===================
test_demo.py::test_demo1
  E:\demo\test_demo.py:7: SyntaxWarning: warning,used to test...
    warnings.warn(SyntaxWarning("warning,used to test..."))

-- Docs: https://docs.pytest.org/en/stable/how-to/capture-warnings.html
=============== 2 passed, 1 warning in 0.02s ===============

(demo-HCIhXOHq) E:\demo>
```

同样，这里也支持当前文件中的所有告警的内容正则匹配，如下所示，从告警信息中匹配 demo 字符的，匹配到的就忽略不显示。

```
import pytest
import warnings
pytestmark = pytest.mark.filterwarnings("ignore:.*demo.*")

def test_demo1():
    print("in test_demo1 ...")
    warnings.warn(SyntaxWarning("warning,used to test..."))
    assert 1 == 1


def test_demo2():
    print("in test_demo2 ...")
    warnings.warn(UserWarning("warning,used to demo..."))
    assert 1 == 1
```

可以看出，此时含有 demo 的 UserWarning 类型的告警已经不显示了，而不含 demo 的 SyntaxWarning 类型的告警依然会显示。

```
(demo-HCIhXOHq) E:\demo> pytest -s
=================== test session starts ===================
platform win32 -- Python 3.7.9, pytest-7.2.0, pluggy-1.0.0
rootdir: E:\demo
plugins: assume-2.4.3, rerunfailures-10.2
collected 2 items

test_demo.py in test_demo1 ...
. in test_demo2 ...
```

```
=================== warnings summary ===================
test_demo.py::test_demo1
  E:\demo\test_demo.py:7: SyntaxWarning: warning,used to test...
    warnings.warn(SyntaxWarning("warning,used to test..."))

-- Docs: https://docs.pytest.org/en/stable/how-to/capture-warnings.html
=============== 2 passed, 1 warning in 0.02s ================

(demo-HCIhX0Hq) E:\demo>
```

10.4 如何关闭所有告警显示

通过命令行参数还可以实现关闭所有告警,人工抛出告警如下所示。

```
import warnings

def test_demo1():
    print("in test_demo1...")
    warnings.warn(SyntaxWarning("warning,used to test..."))
    assert 1 == 1

def test_demo2():
    print("in test_demo2...")
    warnings.warn(UserWarning("warning,used to demo..."))
    assert 1 == 1
```

这里报出了两条告警信息。

```
(demo-HCIhX0Hq) E:\demo> pytest -s
==================== test session starts ====================
platform win32 -- Python 3.7.9, pytest-7.2.0, pluggy-1.0.0
rootdir: E:\demo
plugins: assume-2.4.3, rerunfailures-10.2
collected 2 items

test_demo.py in test_demo1...
.in test_demo2...


=================== warnings summary ===================
```

```
test_demo.py::test_demo1
  E:\demo\test_demo.py:5: SyntaxWarning: warning,used to test...
    warnings.warn(SyntaxWarning("warning,used to test..."))

test_demo.py::test_demo2
  E:\demo\test_demo.py:10: UserWarning: warning,used to demo...
    warnings.warn(UserWarning("warning,used to demo..."))

-- Docs: https://docs.pytest.org/en/stable/how-to/capture-warnings.html
=============== 2 passed, 2 warnings in 0.02s ===============

(demo-HCIhX0Hq) E:\demo>
```

在命令行中使用--disable-warnings参数即可关闭告警显示,这里结果中虽然还显示有2条搞定,但是回显中具体告警内容不再显示了,如下所示。

```
(demo-HCIhX0Hq) E:\demo> pytest --disable-warnings
=================== test session starts ===================
platform win32 -- Python 3.7.9, pytest-7.2.0, pluggy-1.0.0
rootdir: E:\demo
plugins: assume-2.4.3, rerunfailures-10.2
collected 2 items

test_demo.py ..                                       [100%]

=============== 2 passed, 2 warnings in 0.02s ===============

(demo-HCIhX0Hq) E:\demo>
```

通过在命令行中使用如下参数-p no:warnings,可以彻底关闭告警。如下所示,此时结果中已没有告警内容以及告警结果的显示。

```
(demo-HCIhX0Hq) E:\demo> pytest -p no:warnings
=================== test session starts ===================
platform win32 -- Python 3.7.9, pytest-7.2.0, pluggy-1.0.0
rootdir: E:\demo
plugins: assume-2.4.3, rerunfailures-10.2
collected 2 items

test_demo.py ..                                       [100%]

=================== 2 passed in 0.02s ===================

(demo-HCIhX0Hq) E:\demo>
```

10.5 如何通过 Pytest.ini 配置告警或将告警报错

首先,我们看如下测试脚本,在两个测试函数中人为抛出了两个告警。

```python
import warnings

def test_demo1():
    print("in test_demo1 ...")
    warnings.warn(SyntaxWarning("warning,used to test..."))
    assert 1 == 1


def test_demo2():
    print("in test_demo2 ...")
    warnings.warn(UserWarning("warning,used to demo..."))
    assert 1 == 1
```

这里显示出两条告警信息。

```
(demo-HCIhXOHq) E:\demo> pytest -s
==================== test session starts ====================
platform win32 -- Python 3.7.9, pytest-7.2.0, pluggy-1.0.0
rootdir: E:\demo, configfile: pytest.ini
plugins: assume-2.4.3, rerunfailures-10.2
collected 2 items

test_demo.py in test_demo1 ...
.in test_demo2 ...
.

==================== warnings summary ====================
test_demo.py::test_demo1
  E:\demo\test_demo.py:5: SyntaxWarning: warning,used to test...
    warnings.warn(SyntaxWarning("warning,used to test..."))

test_demo.py::test_demo2
  E:\demo\test_demo.py:10: UserWarning: warning,used to demo...
    warnings.warn(UserWarning("warning,used to demo..."))

-- Docs: https://docs.pytest.org/en/stable/how-to/capture-warnings.html
============== 2 passed, 2 warnings in 0.03s ==============
```

```
(demo-HCIhX0Hq) E:\demo>
```

不论通过命令行的方式还是装饰器的方式,其实都可以在 Pytest.ini 中配置文件,如创建 Pytest.ini,编写代码时,即指定 filterwarnings 可将所有的告警忽略,不显示。

```
[pytest]
filterwarnings =
    ignore
```

此时,不再打印告警信息。

```
(demo-HCIhX0Hq) E:\demo> pytest -s
==================== test session starts ====================
platform win32 -- Python 3.7.9, pytest-7.2.0, pluggy-1.0.0
rootdir: E:\demo, configfile: pytest.ini
plugins: assume-2.4.3, rerunfailures-10.2
collected 2 items

test_demo.py in test_demo1 ...
. in test_demo2 ...
.

==================== 2 passed in 0.02s ====================

(demo-HCIhX0Hq) E:\demo>
```

同样,若指定 filterwarnings 为 error,则可将所有的告警转换为报错。

```
[pytest]
filterwarnings =
    error
```

此时,显示有两个失败,失败的原因为代码执行过程中产生了告警。

```
(demo-HCIhX0Hq) E:\demo> pytest -s
==================== test session starts ====================
platform win32 -- Python 3.7.9, pytest-7.2.0, pluggy-1.0.0
rootdir: E:\demo, configfile: pytest.ini
plugins: assume-2.4.3, rerunfailures-10.2
collected 2 items

test_demo.py in test_demo1 ...
Fin test_demo2 ...
F

========================= FAILURES =========================
```

```
_____ test_demo1 _____

    def test_demo1():
        print("in test_demo1 ...")
>       warnings.warn(SyntaxWarning("warning,used to test..."))
E       SyntaxWarning: warning,used to test...

test_demo.py:5: SyntaxWarning
_____ test_demo2 _____

    def test_demo2():
        print("in test_demo2 ...")
>       warnings.warn(UserWarning("warning,used to demo..."))
E       UserWarning: warning,used to demo...

test_demo.py:10: UserWarning
=================== short test summary info ===================
FAILED test_demo.py::test_demo1 - SyntaxWarning: warning,used to test...
FAILED test_demo.py::test_demo2 - UserWarning: warning,used to demo...
===================== 2 failed in 0.08s =====================

(demo-HCIhXOHq) E:\demo>
```

当然,我们也可以指定忽略特定的告警类型,如忽略不显示 UserWarning 类型的告警。

```
[pytest]
filterwarnings =
    ignore::UserWarning
```

可以看出,此时 UserWarning 类型的告警不再显示了。

```
(demo-HCIhXOHq) E:\demo> pytest -s
===================== test session starts =====================
platform win32 -- Python 3.7.9, pytest-7.2.0, pluggy-1.0.0
rootdir: E:\demo, configfile: pytest.ini
plugins: assume-2.4.3, rerunfailures-10.2
collected 2 items

test_demo.py in test_demo1 ...
.in test_demo2 ...
.

===================== warnings summary =====================
```

```
test_demo.py::test_demo1
  E:\demo\test_demo.py:5: SyntaxWarning: warning,used to test...
    warnings.warn(SyntaxWarning("warning,used to test..."))

-- Docs: https://docs.pytest.org/en/stable/how-to/capture-warnings.html
============== 2 passed, 1 warning in 0.02s ===============

(demo-HCIhXOHq) E:\demo>
```

同样，也可以通过配置正则表达式来过滤告警，对匹配的告警进行忽略，不显示处理或报错处理，如下所示即为对告警信息含有 demo 的告警进行忽略不显示。

```
[pytest]
filterwarnings =
    ignore:.*demo.*
```

我们看到，此时含有 demo 的告警不再显示。

```
(demo-HCIhXOHq) E:\demo> pytest -s
==================== test session starts ====================
platform win32 -- Python 3.7.9, pytest-7.2.0, pluggy-1.0.0
rootdir: E:\demo, configfile: pytest.ini
plugins: assume-2.4.3, rerunfailures-10.2
collected 2 items

test_demo.py in test_demo1 ...
. in test_demo2 ...
.

==================== warnings summary ====================
test_demo.py::test_demo1
  E:\demo\test_demo.py:5: SyntaxWarning: warning,used to test...
    warnings.warn(SyntaxWarning("warning,used to test..."))

-- Docs: https://docs.pytest.org/en/stable/how-to/capture-warnings.html
============== 2 passed, 1 warning in 0.02s ===============

(demo-HCIhXOHq) E:\demo>
```

对于命令行中使用的参数，可以在 Pytest.ini 中通过 addopts 指定。如下所示为关闭告警显示。

```
[pytest]
addopts =--disable-warnings
```

执行结果中显示有两条告警，但告警的具体内容不再显示。

```
(demo-HCIhXOHq) E:\demo> pytest -s
==================== test session starts ====================
platform win32 -- Python 3.7.9, pytest-7.2.0, pluggy-1.0.0
rootdir: E:\demo, configfile: pytest.ini
plugins: assume-2.4.3, rerunfailures-10.2
collected 2 items

test_demo.py in test_demo1 ...
. in test_demo2 ...
.

=============== 2 passed, 2 warnings in 0.02s ===============

(demo-HCIhXOHq) E:\demo>
```

通过指定参数,可以彻底关闭告警。

```
[pytest]
addopts = -p no:warnings
```

执行结果中不再显示搞定内容。

```
(demo-HCIhXOHq) E:\demo> pytest
==================== test session starts ====================
platform win32 -- Python 3.7.9, pytest-7.2.0, pluggy-1.0.0
rootdir: E:\demo, configfile: pytest.ini
plugins: assume-2.4.3, rerunfailures-10.2
collected 2 items

test_demo.py ..                                       [100%]

==================== 2 passed in 0.02s ====================

(demo-HCIhXOHq) E:\demo>
```

10.6 如何对产生的告警进行断言

通常,可以使用 with 语句对告警的类型进行断言,即断言代码中会产生 UserWarning 类型的告警。

```
import pytest
import warnings
```

```python
def test_demo():
    print("in test_demo...")
    with pytest.warns(UserWarning):
        warnings.warn(UserWarning("warning,used to demo..."))
```

执行结果显示断言通过。

```
(demo-HCIhXOHq) E:\demo> pytest -s
==================== test session starts ====================
platform win32 -- Python 3.7.9, pytest-7.2.0, pluggy-1.0.0
rootdir: E:\demo, configfile: pytest.ini
plugins: assume-2.4.3, rerunfailures-10.2
collected 1 item

test_demo.py in test_demo...
.

==================== 1 passed in 0.02s ====================

(demo-HCIhXOHq) E:\demo>
```

我们将测试代码修改为如下代码,并在这些代码中抛出 SyntaxWarning 类型的告警,实际上断言期望的是 UserWarning 类型的告警。

```python
import pytest
import warnings


def test_demo():
    print("in test_demo...")
    with pytest.warns(UserWarning):
        warnings.warn(SyntaxWarning("warning,used to demo..."))
```

我们看,执行结果中报错了,显示中并未有 UserWarning 类型的告警产生。

```
(demo-HCIhXOHq) E:\demo> pytest -s
==================== test session starts ====================
platform win32 -- Python 3.7.9, pytest-7.2.0, pluggy-1.0.0
rootdir: E:\demo, configfile: pytest.ini
plugins: assume-2.4.3, rerunfailures-10.2
collected 1 item

test_demo.py in test_demo...
F

========================== FAILURES ==========================
```

```
_____ test_demo _____

    def test_demo():
        print("in test_demo...")
        with pytest.warns(UserWarning):
>           warnings.warn(SyntaxWarning("warning,used to demo..."))
E           Failed: DID NOT WARN. No warnings of type (<class 'UserWarning'>,) were emitted.
E           The list of emitted warnings is: [SyntaxWarning('warning,used to demo...')].

test_demo.py:8: Failed
================== short test summary info ==================
FAILED test_demo.py::test_demo - Failed: DID NOT WARN. No warnings of type (<class 'User...
==================== 1 failed in 0.07s ====================

(demo-HCIhXOHq) E:\demo>
```

此外,还可以通过正则表达式对告警内容进行断言,如下判断告警内容中含有 demo。

```
import pytest
import warnings


def test_demo():
    print("in test_demo...")
    with pytest.warns(UserWarning,match=r'.*demo.*'):
        warnings.warn(UserWarning("warning,used to demo..."))
```

执行结果如下所示,即断言通过。

```
(demo-HCIhXOHq) E:\demo> pytest -s
==================== test session starts ====================
platform win32 -- Python 3.7.9, pytest-7.2.0, pluggy-1.0.0
rootdir: E:\demo, configfile: pytest.ini
plugins: assume-2.4.3, rerunfailures-10.2
collected 1 item

test_demo.py in test_demo...
.
==================== 1 passed in 0.02s ====================

(demo-HCIhXOHq) E:\demo>
```

除此以外,我们还可以对函数调用中的告警类型进行断言,即断言 add 函数中会产生 UserWarning 类型的告警,同时,add 函数需要给定两个参数 a 和 b。

```
import pytest
```

```python
import warnings

def add(a,b):
    print("a:",a)
    print("b:",b)
    warnings.warn(UserWarning("warning,used to demo..."))

def test_demo():
    print("in test_demo...")
    pytest.warns(UserWarning,add,10,20)
```

从执行结果可以看出,参数 a 和 b 能够正常传入,且做到了对函数中产生的告警类型的断言。

```
(demo-HCIhXOHq) E:\demo> pytest -s
==================== test session starts ====================
platform win32 -- Python 3.7.9, pytest-7.2.0, pluggy-1.0.0
rootdir: E:\demo, configfile: pytest.ini
plugins: assume-2.4.3, rerunfailures-10.2
collected 1 item

test_demo.py in test_demo...
a: 10
b: 20
.

==================== 1 passed in 0.02s ====================

(demo-HCIhXOHq) E:\demo>
```

10.7 如何通过 recwarn 记录用例中产生的告警

在 Pytest 框架中,利用 recwarn 这个内置的 fixture,可以对测试用例产生的所有告警进行记录,并可以在最后进行统一的解析处理。这里人为产生了两条不同类型的告警,UserWarning 和 SyntaxWarning 类型的告警。我们在测试代码最后,对当前测试用例中产生的告警条数进行了断言,并且打印了告警,包括类型、告警信息,以及产生告警的文件以及产生告警的代码行数。

```python
import warnings

def test_demo(recwarn):
    warnings.warn(UserWarning("user warning test..."))
    warnings.warn(SyntaxWarning("syntax warning demo..."))
```

```
            assert len(recwarn) == 2
            w = recwarn.pop()
            print(" -------------------------- ")
            print(w.category)
            print(w.message)
            print(w.filename)
            print(w.lineno)
            print(" -------------------------- ")
            w = recwarn.pop()
            print(w.category)
            print(w.message)
            print(w.filename)
            print(w.lineno)
            print(" -------------------------- ")
```

从执行结果可以看出,fixture 解析时是按照从前往后的顺序解析的,可以理解为告警是存放在一个先进先出的队列中。

```
(demo-HCIhXOHq) E:\demo> pytest -s
==================== test session starts ====================
platform win32 -- Python 3.7.9, pytest-7.2.0, pluggy-1.0.0
rootdir: E:\demo, configfile: pytest.ini
plugins: assume-2.4.3, rerunfailures-10.2
collected 1 item

test_demo.py ----------------------------
<class 'UserWarning'>
user warning test ...
E:\demo\test_demo.py
5
 --------------------------
<class 'SyntaxWarning'>
syntax warning demo ...
E:\demo\test_demo.py
6
 --------------------------
.

==================== 1 passed in 0.02s ====================

(demo-HCIhXOHq) E:\demo>
```

有了这个记录告警的功能,其实就可以做很多事情,比如在用例执行后对产生的告警进行处理,又如可以根据告警类型或告警信息关键字加以提取等,对特定的告警进行特殊的处理,以及报错等。

第 11 章　日志和控制台输出管理

11.1　实时标准输出和捕获标准输出

在 Pytest 执行过程中，执行日志显示其实是比较复杂的，我们将详细进行讲解。学习之前，先要理解什么是实时标准输出和捕获标准输出。从字面意思来理解，实时标准输出就是在脚本执行过程中打印的标准输出，而捕获标准输出，就是在执行的过程中被后台程序捕获的，但并未在控制台打印的标准输出。Pytest 对捕获标准输出的处理方式是当测试用例成功时，捕获标准输出不会打印，而当测试用例失败时，才会将捕获标准输出做处理。

将关于后续实时日志和捕获日志，这里要特别说明这里的标准输出主要是指 stdout 中的内容可理解为通过 print 语句打印的内容。

我们看如下一段测试代码，这里在测试脚本中打印了一句话，然后测试断言是成功的。

```
def test_demo():
    print("in test_demo...")
    assert 1 == 1
```

使用 Pytest 命令执行，需要说明 print 语句打印的内容要么在实时标准输出中，要么在捕获标准输出中，即不会同时在实时标准输出和捕获标准输出中，这里因为用例通过时，是不会显示捕获标准输出的，因此这里只显示了实时标准输出。但是从打印的记录可以看出，这里并未打印测试脚本中打印的那一行话，因为在默认情况下，捕获实时输出是开启的，也就是当用 Pytest 命令执行时，print 的内容是自动打印到捕获标准输出的。

```
(demo-HCIhXOHq) E:\demo> pytest
==================== test session starts ====================
platform win32 -- Python 3.7.9, pytest-7.2.0, pluggy-1.0.0
rootdir: E:\demo
plugins: assume-2.4.3, rerunfailures-10.2
collected 1 item

test_demo.py .                                         [100%]
```

```
================== 1 passed in 0.02s ==================
```

(demo - HCIhX0Hq) E:\demo>

为了能显示出捕获标准输出,我们可以将测试脚本中的断言改为失败,如下所示。

```
def test_demo():
    print("in test_demo ...")
    assert 1 == 2
```

执行结果如下所示,Captured stdout call 下面的一行内容即为捕获标准输出的内容,这里显示输出内容打印出来了。

```
(demo - HCIhX0Hq) E:\demo> pytest
================== test session starts ==================
platform win32 -- Python 3.7.9, pytest - 7.2.0, pluggy - 1.0.0
rootdir: E:\demo
plugins: assume - 2.4.3, rerunfailures - 10.2
collected 1 item

test_demo.py F                                    [100%]

========================= FAILURES =========================
_____ test_demo _____

    def test_demo():
        print("in test_demo ...")
>       assert 1 == 2
E       assert 1 == 2

test_demo.py:4: AssertionError
-------------------- Captured stdout call --------------------
in test_demo ...
================ short test summary info ================
FAILED test_demo.py::test_demo - assert 1 == 2
================== 1 failed in 0.07s ==================
```

(demo - HCIhX0Hq) E:\demo>

这就是实时标准输出和捕获标准输出,根据 Pytest 的设计原理,在默认情况下,捕获标准输出是开启的,当执行一批自动化脚本的时候,脚本中通过 print 打印的信息都会记录到捕获标准输出中,而当用例失败时,会在执行完成脚本之后都转交给捕获标准输出,打印出来,这样就极大方便了定位。

11.2 如何打开或关闭实时输出和捕获标准输出

我们知道,Pytest 框架在默认情况下是打开捕获标准输出的,即当用例断言失败时,在默认情况下 Pytest 会自动将脚本中的 print 的内容放在捕获标准输出中打印出来。我们还以如下最简单的测试脚本为例进行说明。注意这里要让断言失败,因为如果断言成功了,则 Pytest 的机制就不会显示捕获标准输出了。

```
def test_demo():
    print("in test_demo...")
    assert 1 == 2
```

可以看到,此时在 Captured stdout call 捕获标准输出中将脚本中打印的信息打印出来了。这说明在默认情况下,Pytest 是开启了捕获标准输出的。

```
(demo-HCIhXOHq) E:\demo> pytest
==================== test session starts ====================
platform win32 -- Python 3.7.9, pytest-7.2.0, pluggy-1.0.0
rootdir: E:\demo
plugins: assume-2.4.3, rerunfailures-10.2
collected 1 item

test_demo.py F                                         [100%]

========================== FAILURES ==========================
_____ test_demo _____

    def test_demo():
        print("in test_demo...")
>       assert 1 == 2
E       assert 1 == 2

test_demo.py:4: AssertionError
-------------------- Captured stdout call --------------------
in test_demo...
==================== short test summary info ====================
FAILED test_demo.py::test_demo - assert 1 == 2
===================== 1 failed in 0.07s =====================

(demo-HCIhXOHq) E:\demo>
```

当然,开启捕获标准输出时,脚本中的打印的信息就不会在实时标准输出中显示。

我们可以将脚本修改为断言成功,此时 Pytest 的机制是不会打印捕获标准输出,但是会打印实时标准输出。

```python
def test_demo():
    print("in test_demo ...")
    assert 1 == 1
```

执行结果如下所示,可以看出,此时实时标准输出中并未把脚本中的 print 的信息打印出来。这与上面的分析结论是一致的,即 Pytest 默认情况下开启了捕获标准输出,而此时用例断言成功,因此不会显示捕获标准输出,控制台输出中也没有显示脚本中的 print 的信息。

```
(demo-HCIhXOHq) E:\demo> pytest
==================== test session starts ====================
platform win32 -- Python 3.7.9, pytest-7.2.0, pluggy-1.0.0
rootdir: E:\demo
plugins: assume-2.4.3, rerunfailures-10.2
collected 1 item

test_demo.py .                                       [100%]

==================== 1 passed in 0.02s ====================

(demo-HCIhXOHq) E:\demo>
```

Pytest 提供了 -s 的参数用来关闭捕获标准输出。当关闭捕获标准输出时,脚本中的 print 的内容就会在实时标准输出中打印出来,当使用 Pytest-s 命令执行时,断言此时仍然是成功的,我们可以从控制台的输出中看到此时脚本中的 print 的内容已经显示出来了。

```
(demo-HCIhXOHq) E:\demo> pytest -s
==================== test session starts ====================
platform win32 -- Python 3.7.9, pytest-7.2.0, pluggy-1.0.0
rootdir: E:\demo
plugins: assume-2.4.3, rerunfailures-10.2
collected 1 item

test_demo.py in test_demo ...
.

==================== 1 passed in 0.02s ====================

(demo-HCIhXOHq) E:\demo>
```

为了进一步验证-s参数是关闭捕获标准输出的结论，我们将脚本修改为断言失败，因为在断言失败的情况下，如果没有关闭捕获标准输出，就会看到捕获标准输出的打印，而如果真的关闭了，即使脚本失败了也不会显示捕获标准输出了。脚本修改为如下所示。

```
def test_demo():
    print("in test_demo ...")
    assert 1 == 2
```

此时，使用Pytest-s命令执行的结果如下所示，可以看出，此时控制台的输出中没有了Captured stdout call的内容，这就验证了上面介绍的结论。

```
(demo-HCIhXOHq) E:\demo> pytest -s
==================== test session starts ====================
platform win32 -- Python 3.7.9, pytest-7.2.0, pluggy-1.0.0
rootdir: E:\demo
plugins: assume-2.4.3, rerunfailures-10.2
collected 1 item

test_demo.py in test_demo ...
F

========================= FAILURES =========================
_____ test_demo _____

    def test_demo():
        print("in test_demo ...")
>       assert 1 == 2
E       assert 1 == 2

test_demo.py:4: AssertionError
================= short test summary info =================
FAILED test_demo.py::test_demo - assert 1 == 2
==================== 1 failed in 0.07s ====================

(demo-HCIhXOHq) E:\demo>
```

因此，总结一下，Pytest框架是通过-s参数关闭捕获标准输出的，在许多教程或者书籍中，有不少说法讲到-s参数是用来显示详细信息的，这是一种简单的假想，不太准确的。实际上-s是关闭捕获标准输出的。

11.3 如何使用logging模块写日志

logging是Python语言中的一个专门用来写日志的日志模块。日志级别通常分为

debug、info、warning、error、critical 几个级别。一般情况下，默认的日志级别为 warning，在调试或者测试阶段，日志级别可以设置为 debug 或者 info，当在生产环境上线后，日志级别一般为 warning 或者 error 级别。下面我们就快速体验一下 logging 模块写日志的用法，这里先创建一个 Python 的 demo.py 文件，然后在 Python 文件中使用 logging 写日志，如使用 logging 对每一个日志级别分别写一条日志，如下所示。

```
import logging

logging.debug("this is debug log")
logging.info("this is info log")
logging.warning("this is warning log")
logging.error("this is error log")
logging.critical("this is critical log")
```

可以看出，此时仅打印了 warning、error、critical 级别的日志，这是因为 Python 中默认的级别是 warning 级别，所以低于 warning 级别的日志都不打印了。

```
(demo-HCIhX0Hq) E:\demo> python demo.py
WARNING:root:this is warning log
ERROR:root:this is error log
CRITICAL:root:this is critical log

(demo-HCIhX0Hq) E:\demo>
```

当然，在代码中是可以修改日志级别的，如下代码即是将日志级别修改为了 debug 级别。

```
import logging

logging.basicConfig(level = logging.DEBUG)
logging.debug("this is debug log")
logging.info("this is info log")
logging.warning("this is warning log")
logging.error("this is error log")
logging.critical("this is critical log")
```

再次执行 demo.py 文件，可以看出，此时 debug 和 info 级别的日志都显示出来了。

```
(demo-HCIhX0Hq) E:\demo> python demo.py
DEBUG:root:this is debug log
INFO:root:this is info log
WARNING:root:this is warning log
ERROR:root:this is error log
CRITICAL:root:this is critical log
```

```
(demo-HCIhXOHq) E:\demo>
```

利用 logging 在日志文件中写日志也是很容易的。我们看如下代码，这里配置一个文件即可，同时还可以设置日志的级别，这里仍然设置为 debug 级别，即 debug 及其以上级别的日志均会写入日志文件。

```
import logging

logging.basicConfig(filename = "demo.log", level = logging.DEBUG)
logging.debug("this is debug log")
logging.info("this is info log")
logging.warning("this is warning log")
logging.error("this is error log")
logging.critical("this is critical log")
```

此时，再次执行 demo.py 文件，可以发现，控制台并没有打印，而是在当前目录下生成了一个 demo.log 文件，内容如下。

```
DEBUG:root:this is debug log
INFO:root:this is info log
WARNING:root:this is warning log
ERROR:root:this is error log
CRITICAL:root:this is critical log
```

我们看到，默认情况下写的日志是按照追加的模式，比如再次执行一次，则 demo.log 中的内容变为如下内容。

```
DEBUG:root:this is debug log
INFO:root:this is info log
WARNING:root:this is warning log
ERROR:root:this is error log
CRITICAL:root:this is critical log
DEBUG:root:this is debug log
INFO:root:this is info log
WARNING:root:this is warning log
ERROR:root:this is error log
CRITICAL:root:this is critical log
```

当然，我们可以设置写入日志的模式，比如将如下的 filemode 模式设置为 w，表示每次清空文件再写日志。当然如果把 filemode 设置为 a 则为追加模式，如果不设置，默认情况下也是追加模式。

```
import logging

logging.basicConfig(filename = "demo.log", filemode = 'w', level = logging.DEBUG)
```

```
logging.debug("this is debug log")
logging.info("this is info log")
logging.warning("this is warning log")
logging.error("this is error log")
logging.critical("this is critical log")
```

此时,再次执行 demo.py 文件,因为模式设置为 w 了,因此 demo.log 内容会先清空再写入,内容如下所示。

```
DEBUG:root:this is debug log
INFO:root:this is info log
WARNING:root:this is warning log
ERROR:root:this is error log
CRITICAL:root:this is critical log
```

平常,我们在查看其他产品的日志时,都是会显示文件、时间、代码行数等信息,我们这里也是可以设置的。比如这里直接为设置日志格式,并直接在控制台打印了。

```
import logging

logging.basicConfig(format = ("%(asctime)s | %(levelname)s | %(filename)s:%(lineno)s | %(message)s"),
                    datefmt = "%Y-%m-%d_%H:%M:%S",
                    level = logging.DEBUG)
logging.debug("this is debug log")
logging.info("this is info log")
logging.warning("this is warning log")
logging.error("this is error log")
logging.critical("this is critical log")
```

从执行结果可以看到,此时日志中有时间戳、日志级别、代码文件、代码函数,日志内容等。这个格式基本就是我们希望要的日志格式。

```
(demo-HCIhX0Hq) E:\demo> python demo.py
2022-12-04_22:47:14 | DEBUG | demo.py:6 | this is debug log
2022-12-04_22:47:14 | INFO | demo.py:7 | this is info log
2022-12-04_22:47:14 | WARNING | demo.py:8 | this is warning log
2022-12-04_22:47:14 | ERROR | demo.py:9 | this is error log
2022-12-04_22:47:14 | CRITICAL | demo.py:10 | this is critical log

(demo-HCIhX0Hq) E:\demo>
```

当然,Python 框架中的 logging 模块还有许多其他高级的应用,在 Pytest 框架中只需要这么简单地用 logging 即可,因此这里就不再深入介绍 logging 相关知识了。

11.4 什么是实时日志和捕获日志

实时日志和捕获日志与实时标准输出和捕获标准输出是类似的,其中,实时日志和捕获日志都是用来显示通过 logging 模块打印的日志内容,而 print 打印内容不会在实时日志和捕获日志中显示的。

实时日志顾名思义就是实时显示的日志,而捕获日志就是在执行的过程中将 logging 打印的日志都收集起来,同样 Pytest 框架对捕获日志的显示机制是当测试脚本失败后才会显示,当用例通过时,则不会显示捕获日志。这与实际情况一致,因为在自动化测试过程中,当用例都通过后,我们一般是不关心日志打印的,只有当用例失败后,我们定位失败原因时才会关注出错的信息以及帮助定位的日志信息。有一点需注意,在 Pytest 框架中,实时日志在默认情况下是不显示的,如果要显示实时日志需要主动配置开启实时日志。

下面我们观察一下如下所示的代码,在测试脚本中通过 logging 打印了几条日志信息,用例断言设置通过。

```
import logging

def test_demo():
    logging.warning("this is warning log...")
    logging.error("this is error log...")
    logging.critical("this is critical log...")
    assert 1 == 1
```

执行结果如下所示,可以发现,虽然此时日志打印的是 warning、error 和 critical 日志,按正常来说应该都是可以显示出来的,但是这里却没显示,这就是说在 Pytest 中默认情况下实时日志是关闭的。

```
(demo-HCIhXOHq) E:\demo> pytest
==================== test session starts ====================
platform win32 -- Python 3.7.9, pytest-7.2.0, pluggy-1.0.0
rootdir: E:\demo
plugins: assume-2.4.3, rerunfailures-10.2
collected 1 item

test_demo.py .                                        [100%]

==================== 1 passed in 0.02s ====================

(demo-HCIhXOHq) E:\demo>
```

我们可以先通过 Pytest-o log_cli=true 执行，观察结果如下所示，此时，可以看到 live log call 已经存在了，这就是实时日志。

```
(demo-HCIhXOHq) E:\demo> pytest -o log_cli=true
==================== test session starts ====================
platform win32 -- Python 3.7.9, pytest-7.2.0, pluggy-1.0.0
rootdir: E:\demo
plugins: assume-2.4.3, rerunfailures-10.2
collected 1 item


test_demo.py::test_demo
---------------------- live log call ----------------------
WARNING  root:test_demo.py:4 this is warning log ...
ERROR    root:test_demo.py:5 this is error log ...
CRITICAL root:test_demo.py:6 this is critical log ...
PASSED                                            [100%]


==================== 1 passed in 0.02s ====================

(demo-HCIhXOHq) E:\demo>
```

下面将测试用例断言设置为失败。

```
import logging


def test_demo():
    logging.warning("this is warning log ...")
    logging.error("this is error log ...")
    logging.critical("this is critical log ...")
    assert 1 == 2
```

可以发现，live log call 即实时日志会在执行用例时直接实时显示出来，而在用例执行失败后，会有 Captured log call 的部分，这就是捕获日志，即将用例中通过 logging 打印的日志信息收集起来，待用例执行完成后再打印出来。

```
(demo-HCIhXOHq) E:\demo> pytest -o log_cli=true
==================== test session starts ====================
platform win32 -- Python 3.7.9, pytest-7.2.0, pluggy-1.0.0
rootdir: E:\demo
plugins: assume-2.4.3, rerunfailures-10.2
collected 1 item


test_demo.py::test_demo
---------------------- live log call ----------------------
```

```
WARNING   root:test_demo.py:4 this is warning log ...
ERROR     root:test_demo.py:5 this is error log ...
CRITICAL  root:test_demo.py:6 this is critical log ...
FAILED                                                          [100%]


========================= FAILURES =========================
_____ test_demo _____

    def test_demo():
        logging.warning("this is warning log ...")
        logging.error("this is error log ...")
        logging.critical("this is critical log ...")
>       assert 1 == 2
E       assert 1 == 2

test_demo.py:7: AssertionError
--------------------- Captured log call ---------------------
WARNING   root:test_demo.py:4 this is warning log ...
ERROR     root:test_demo.py:5 this is error log ...
CRITICAL  root:test_demo.py:6 this is critical log ...
================== short test summary info ==================
FAILED test_demo.py::test_demo - assert 1 == 2
==================== 1 failed in 0.08s ====================

(demo-HCIhXOHq) E:\demo>
```

上面即通过实例展示了实时日志与捕获日志的情况。

11.5 如何打开或关闭实时日志和捕获日志

Pytest 框架中默认情况下实时日志是关闭的，捕获日志是打开的，因此这里主要介绍如何打开实时日志以及关闭捕获日志。

我们看如下脚本。

```
import logging

def test_demo():
    logging.warning("this is warning log ...")
    logging.error("this is error log ...")
    logging.critical("this is critical log ...")
    assert 1 == 1
```

开启实时日志有两种方式,一种是通过命令行 Pytest-o log_cli=true 开启,如下所示。

```
(demo-HCIhX0Hq) E:\demo> pytest -o log_cli=true
==================== test session starts ====================
platform win32 -- Python 3.7.9, pytest-7.2.0, pluggy-1.0.0
rootdir: E:\demo
plugins: assume-2.4.3, rerunfailures-10.2
collected 1 item

test_demo.py::test_demo
---------------------- live log call ----------------------
WARNING  root:test_demo.py:4 this is warning log...
ERROR    root:test_demo.py:5 this is error log...
CRITICAL root:test_demo.py:6 this is critical log...
PASSED                                              [100%]

==================== 1 passed in 0.02s ====================

(demo-HCIhX0Hq) E:\demo>
```

在命令行中打开实时日志是比较烦琐的。另外一种日志开启方式就是在 Pytest.ini 文件中做配置来完成。首先要创建 Pytest.ini 文件,然后写入配置,可打开实时日志。

```
[pytest]
log_cli = True
```

此时,直接使用 Pytest 命令即可显示实时日志了,如下所示。

```
(demo-HCIhX0Hq) E:\demo> pytest
==================== test session starts ====================
platform win32 -- Python 3.7.9, pytest-7.2.0, pluggy-1.0.0
rootdir: E:\demo, configfile: pytest.ini
plugins: assume-2.4.3, rerunfailures-10.2
collected 1 item

test_demo.py::test_demo
---------------------- live log call ----------------------
WARNING  root:test_demo.py:4 this is warning log...
ERROR    root:test_demo.py:5 this is error log...
CRITICAL root:test_demo.py:6 this is critical log...
PASSED                                              [100%]
```

```
==================1 passed in 0.02s ==================
```

(demo-HCIhXOHq) E:\demo>

当不设置日志级别时，会默认也是 warning，即 warning 及以上的级别日志才会显示，而 info 和 debug 日志在默认时是不会显示的。

```python
import logging

def test_demo():
    logging.debug("this is debug log ...")
    logging.info("this is info log ...")
    logging.warning("this is warning log ...")
    logging.error("this is error log ...")
    logging.critical("this is critical log ...")
    assert 1 == 1
```

从上述执行结果中可以看到，此时 info 和 debug 日志未显示。

```
(demo-HCIhXOHq) E:\demo> pytest
================== test session starts ==================
platform win32 -- Python 3.7.9, pytest-7.2.0, pluggy-1.0.0
rootdir: E:\demo, configfile: pytest.ini
plugins: assume-2.4.3, rerunfailures-10.2
collected 1 item

test_demo.py::test_demo
---------------------- live log call ----------------------
WARNING  root:test_demo.py:6 this is warning log ...
ERROR    root:test_demo.py:7 this is error log ...
CRITICAL root:test_demo.py:8 this is critical log ...
PASSED                                              [100%]

==================1 passed in 0.02s ==================
```

(demo-HCIhXOHq) E:\demo>

在 Pytest.ini 中开启实时日志时，可以将日志级别以及日志格式同步设置一下，Pytest.ini 设置如下所示。

```ini
[pytest]
log_cli = True
log_cli_level = debug
log_cli_format = %(asctime)s %(levelname)s %(message)s
log_cli_date_format = %Y-%m-%d %H:%M:%S
```

此时,再次执行,debug 和 info 日志此时均显示出来了,而且每行日志均带有时间戳、日志级别、日志内容。

```
(demo-HCIhXOHq) E:\demo> pytest
==================== test session starts ====================
platform win32 -- Python 3.7.9, pytest-7.2.0, pluggy-1.0.0
rootdir: E:\demo, configfile: pytest.ini
plugins: assume-2.4.3, rerunfailures-10.2
collected 1 item

test_demo.py::test_demo
---------------------- live log call ----------------------
2022-12-05 16:22:17 DEBUG this is debug log...
2022-12-05 16:22:17 INFO this is info log...
2022-12-05 16:22:17 WARNING this is warning log...
2022-12-05 16:22:17 ERROR this is error log...
2022-12-05 16:22:17 CRITICAL this is critical log...
PASSED                                              [100%]

==================== 1 passed in 0.02s ====================

(demo-HCIhXOHq) E:\demo>
```

捕获日志在默认情况下是开启的,即当用例失败时就会自动显示捕获的日志,如将测试用例修改为失败,如下所示。

```
import logging

def test_demo():
    logging.debug("this is debug log...")
    logging.info("this is info log...")
    logging.warning("this is warning log...")
    logging.error("this is error log...")
    logging.critical("this is critical log...")
    assert 1 == 2
```

从执行结果中看出,用例失败后显示 Captured log call 了捕获日志。

```
(demo-HCIhXOHq) E:\demo> pytest
==================== test session starts ====================
platform win32 -- Python 3.7.9, pytest-7.2.0, pluggy-1.0.0
rootdir: E:\demo, configfile: pytest.ini
plugins: assume-2.4.3, rerunfailures-10.2
collected 1 item
```

```
test_demo.py::test_demo
------------------------ live log call ------------------------
2022-12-05 16:28:59 DEBUG this is debug log...
2022-12-05 16:28:59 INFO this is info log...
2022-12-05 16:28:59 WARNING this is warning log...
2022-12-05 16:28:59 ERROR this is error log...
2022-12-05 16:28:59 CRITICAL this is critical log...
FAILED                                                    [100%]

========================= FAILURES =========================
_____ test_demo _____

    def test_demo():
        logging.debug("this is debug log...")
        logging.info("this is info log...")
        logging.warning("this is warning log...")
        logging.error("this is error log...")
        logging.critical("this is critical log...")
>       assert 1 == 2
E       assert 1 == 2

test_demo.py:9: AssertionError
--------------------- Captured log call ---------------------
DEBUG    root:test_demo.py:4 this is debug log...
INFO     root:test_demo.py:5 this is info log...
WARNING  root:test_demo.py:6 this is warning log...
ERROR    root:test_demo.py:7 this is error log...
CRITICAL root:test_demo.py:8 this is critical log...
================= short test summary info =================
FAILED test_demo.py::test_demo - assert 1 == 2
==================== 1 failed in 0.08s ====================

(demo-HCIhXOHq) E:\demo>
```

捕获日志 Pytest 默认是打开的，这里顺便也提一下，对于捕获日志，也可以配置日志级别、时间戳、格式等，但捕获日志配置与实时日志配置关键字是不一样的，如下所示即为配置捕获日志格式的。

```
[pytest]
log_level = debug
log_format = %(asctime)s %(levelname)s %(message)s
log_date_format = %Y-%m-%d %H:%M:%S
```

可以看到，此时捕获日志已经将 debug 和 info 级别的日志显示出来了，并且也带有时间戳和日志级别的信息。

```
(demo-HCIhXOHq) E:\demo> pytest
===================== test session starts =====================
platform win32 -- Python 3.7.9, pytest-7.2.0, pluggy-1.0.0
rootdir: E:\demo, configfile: pytest.ini
plugins: assume-2.4.3, rerunfailures-10.2
collected 1 item

test_demo.py F                                          [100%]

========================== FAILURES ===========================
_____ test_demo _____

    def test_demo():
        logging.debug("this is debug log ...")
        logging.info("this is info log ...")
        logging.warning("this is warning log ...")
        logging.error("this is error log ...")
        logging.critical("this is critical log ...")
>       assert 1 == 2
E       assert 1 == 2

test_demo.py:9: AssertionError
--------------------- Captured log call -----------------------
2022-12-05 17:12:33 DEBUG this is debug log ...
2022-12-05 17:12:33 INFO this is info log ...
2022-12-05 17:12:33 WARNING this is warning log ...
2022-12-05 17:12:33 ERROR this is error log ...
2022-12-05 17:12:33 CRITICAL this is critical log ...
================== short test summary info ===================
FAILED test_demo.py::test_demo - assert 1 == 2
===================== 1 failed in 0.08s ======================

(demo-HCIhXOHq) E:\demo>
```

而若想关闭捕获日志,只需要使用--show-capture=no参数即可完成,我们看如下代码,此时没有捕获日志显示了。在实际应用中,一般不需要将捕获日志关闭,因为在默认用例通过时,捕获日志就不显示,而当用例失败,正需要查看捕获日志,因此,对于捕获日志一般使其默认打开即可,需要做的是设置日志打印的格式等。

```
(demo-HCIhXOHq) E:\demo> pytest --show-capture=no
===================== test session starts =====================
platform win32 -- Python 3.7.9, pytest-7.2.0, pluggy-1.0.0
```

```
rootdir: E:\demo, configfile: pytest.ini
plugins: assume-2.4.3, rerunfailures-10.2
collected 1 item

test_demo.py F                                              [100%]

========================== FAILURES ==========================
_____ test_demo _____

    def test_demo():
        logging.debug("this is debug log ...")
        logging.info("this is info log ...")
        logging.warning("this is warning log ...")
        logging.error("this is error log ...")
        logging.critical("this is critical log ...")
>       assert 1 == 2
E       assert 1 == 2

test_demo.py:9: AssertionError
================== short test summary info ==================
FAILED test_demo.py::test_demo - assert 1 == 2
==================== 1 failed in 0.07s ====================

(demo-HCIhXOHq) E:\demo>
```

11.6 caplog 的应用场景及使用方法

caplog 是 Pytest 框架一个内置的 fixture,用于处理日志操作。

11.6.1 如何在测试用例中设置日志级别

通过 caplog 可以在特定的测试函数内设置日志级别,而不影响其全局的日志级别,我们首先在 Pytest.ini 中开启实时日志。

```
[pytest]
log_cli = True
```

我们在测试代码 test_demo 和 test_demo2 时均打印 debug、info、warning、error、critical 级别的日志,在 test_demo2 中,使用 caplog 临时将日志级别设置为 debug 级别,因为在默认情况下,日志级别为 warning,因此,根据理论分析,test_demo 中的日志打印将采用默认的日志级别,即只会显示 warning、error、critical 级别的日志,而在 test_demo2 中,由于临时将日志级别设置为 debug,因此这里所有的日志都将显示已处理。

```python
import logging

def test_demo():
    logging.debug("this is debug log ...")
    logging.info("this is info log ...")
    logging.warning("this is warning log ...")
    logging.error("this is error log ...")
    logging.critical("this is critical log ...")
    assert 1 == 1

def test_demo2(caplog):
    caplog.set_level(logging.DEBUG)
    logging.debug("this is debug log ...")
    logging.info("this is info log ...")
    logging.warning("this is warning log ...")
    logging.error("this is error log ...")
    logging.critical("this is critical log ...")
    assert 1 == 1
```

可以看出，在 test_demo 中只显示了 warning、error、critical 级别的日志，而在 test_demo2 中所有的日志都显示出来了，即此时的日志级别已经通过 caplog 修改为 debug 级别了。

```
(demo-HCIhXOHq) E:\demo> pytest
==================== test session starts ====================
platform win32 -- Python 3.7.9, pytest-7.2.0, pluggy-1.0.0
rootdir: E:\demo, configfile: pytest.ini
plugins: assume-2.4.3, rerunfailures-10.2
collected 2 items


test_demo.py::test_demo
---------------------- live log call ----------------------
WARNING  root:test_demo.py:6 this is warning log ...
ERROR    root:test_demo.py:7 this is error log ...
CRITICAL root:test_demo.py:8 this is critical log ...
PASSED                                              [ 50%]
test_demo.py::test_demo2
---------------------- live log call ----------------------
DEBUG    root:test_demo.py:13 this is debug log ...
INFO     root:test_demo.py:14 this is info log ...
WARNING  root:test_demo.py:15 this is warning log ...
ERROR    root:test_demo.py:16 this is error log ...
CRITICAL root:test_demo.py:17 this is critical log ...
```

PASSED   [100%]

================== 2 passed in 0.03s ==================

(demo-HCIhXOHq) E:\demo>

11.6.2 如何对日志级别进行断言

caplog 会将日志都记录在 records 属性中,这样就可以在测试脚本末尾,对当前测试脚本中产生的日志级别进行判断,如脚本中可能存在某些条件触发时写入 error 的日志,而在脚本末尾则可以对日志级别进行断言。要求日志不能有 error 日志,可以使用类似如下所示的测试代码完成。

```python
import logging

def test_demo(caplog):
    logging.warning("this is warning log...")
    logging.error("this is error log...")
    logging.critical("this is critical log...")
    for record in caplog.records:
        assert record.levelname != "ERROR"
```

执行结果如下所示。

```
(demo-HCIhXOHq) E:\demo> pytest
==================== test session starts ====================
platform win32 -- Python 3.7.9, pytest-7.2.0, pluggy-1.0.0
rootdir: E:\demo, configfile: pytest.ini
plugins: assume-2.4.3, rerunfailures-10.2
collected 1 item

test_demo.py::test_demo
---------------------- live log call ----------------------
WARNING  root:test_demo.py:4 this is warning log...
ERROR    root:test_demo.py:5 this is error log...
CRITICAL root:test_demo.py:6 this is critical log...
FAILED                                                   [100%]

========================= FAILURES =========================
_____ test_demo _____

caplog = <_pytest.logging.LogCaptureFixture object at 0x0000029BEF29C608>

    def test_demo(caplog):
```

```
            logging.warning("this is warning log...")
            logging.error("this is error log...")
            logging.critical("this is critical log...")
            for record in caplog.records:
>               assert record.levelname != "ERROR"
E               assert 'ERROR' != 'ERROR'
E                + where 'ERROR' = <LogRecord: root, 40, E:\demo\test_demo.py, 5, "this
is error log...">.levelname

test_demo.py:8: AssertionError
--------------------- Captured log call ---------------------
WARNING  root:test_demo.py:4 this is warning log...
ERROR    root:test_demo.py:5 this is error log...
CRITICAL root:test_demo.py:6 this is critical log...
================= short test summary info =================
FAILED test_demo.py::test_demo - assert 'ERROR' != 'ERROR'
==================== 1 failed in 0.08s ====================

(demo-HCIhX0Hq) E:\demo>
```

11.6.3　如何对日志内容进行断言

caplog 同样可以做到对日志的内容进行断言，判断日志中是否有 error log 内容，通过 record 的 message 即可获得日志的内容。

```
import logging

def test_demo(caplog):
    logging.warning("this is warning log...")
    logging.error("this is error log...")
    logging.critical("this is critical log...")
    for record in caplog.records:
        assert "error log" not in record.meesge
```

可以看到，断言错误了，即日志中含有 error log 内容。

```
(demo-HCIhX0Hq) E:\demo> pytest -s
==================== test session starts ====================
platform win32 -- Python 3.7.9, pytest-7.2.0, pluggy-1.0.0
rootdir: E:\demo, configfile: pytest.ini
plugins: assume-2.4.3, rerunfailures-10.2
collected 1 item


test_demo.py::test_demo
```

```
---------------------- live log call ----------------------
WARNING  root:test_demo.py:4 this is warning log...
ERROR    root:test_demo.py:5 this is error log...
CRITICAL root:test_demo.py:6 this is critical log...
FAILED


========================= FAILURES =========================
_____ test_demo _____

caplog = <_pytest.logging.LogCaptureFixture object at 0x00000260C060D6C8>

    def test_demo(caplog):
        logging.warning("this is warning log...")
        logging.error("this is error log...")
        logging.critical("this is critical log...")
        for record in caplog.records:
>           assert "error log" not in record.meesge
E           AttributeError: 'LogRecord' object has no attribute 'meesge'

test_demo.py:8: AttributeError
-------------------- Captured log call --------------------
WARNING  root:test_demo.py:4 this is warning log...
ERROR    root:test_demo.py:5 this is error log...
CRITICAL root:test_demo.py:6 this is critical log...
================= short test summary info =================
FAILED test_demo.py::test_demo - AttributeError: 'LogRecord' object has no attribute 'me...
=================== 1 failed in 0.08s ===================

(demo-HCIhXOHq) E:\demo>
```

11.6.4　如何对日志级别和日志内容同时进行断言

caplog 还可以对 logger、日志级别、日志内容组成的元组进行判决，如 caplog.record_tuples 保存了所有 logger、日志级别、日志内容组成的元组集合。

我们看如下测试代码，有一点需要注意，当直接使用 logging.xxx 打印日志时，实际上使用的 logger 是 root，如要定义自己的 logger，在此处就会很容易理解。

```
import logging

def test_demo(caplog):
    logging.warning("this is warning log...")
    logging.error("this is error log...")
```

```
logging.critical("this is critical log...")
assert ("root",logging.ERROR,"this is error log...") in caplog.record_tuples
```

可以看到，此时断言成功了。

```
(demo-HCIhX0Hq) E:\demo> pytest
===================== test session starts =====================
platform win32 -- Python 3.7.9, pytest-7.2.0, pluggy-1.0.0
rootdir: E:\demo, configfile: pytest.ini
plugins: assume-2.4.3, rerunfailures-10.2
collected 1 item

test_demo.py::test_demo
---------------------- live log call ----------------------
WARNING  root:test_demo.py:4 this is warning log...
ERROR    root:test_demo.py:5 this is error log...
CRITICAL root:test_demo.py:6 this is critical log...
PASSED                                                [100%]

===================== 1 passed in 0.02s =====================

(demo-HCIhX0Hq) E:\demo>
```

11.6.5　在测试用例中如何获取 setup 中的日志

在有些测试场景下，我们希望在测试步骤中对 setup 步骤的日志进行一定的处理，比如判断 setup 的日志中是否有 ERROR 级别的日志，那么 caplog 就可以提供这样的功能。

我们在 function_fixture 中的关键字 yield 之前写了几条日志信息，然后在测试脚本中对 setup 中的日志进行判断，如下所示。

```python
import logging
import pytest

@pytest.fixture(scope="function", autouse=True)
def function_fixture():
    logging.warning("this is warning log...")
    logging.error("this is error log...")
    logging.critical("this is critical log...")
    yield

def test_demo(caplog):
    setup_records = caplog.get_records("setup")
    for record in setup_records:
```

```
            assert record.levelname != "ERROR"
```

可以看到,此时已经获取了 setup 中的日志,并且断言失败。

```
(demo-HCIhXOHq) E:\demo> pytest
========================= test session starts =========================
platform win32 -- Python 3.7.9, pytest-7.2.0, pluggy-1.0.0
rootdir: E:\demo, configfile: pytest.ini
plugins: assume-2.4.3, rerunfailures-10.2
collected 1 item


test_demo.py::test_demo
----------------------- live log setup -----------------------
WARNING  root:test_demo.py:6 this is warning log...
ERROR    root:test_demo.py:7 this is error log...
CRITICAL root:test_demo.py:8 this is critical log...
FAILED                                                   [100%]


=========================== FAILURES ===========================
_____ test_demo _____

caplog = <_pytest.logging.LogCaptureFixture object at 0x00000239EF30D188>

    def test_demo(caplog):
        setup_records = caplog.get_records("setup")
        for record in setup_records:
>           assert record.levelname != "ERROR"
E           assert 'ERROR' != 'ERROR'
E            +  where 'ERROR' = <LogRecord: root, 40, E:\demo\test_demo.py, 7, "this is error log...">.levelname

test_demo.py:14: AssertionError
-------------------- Captured log setup --------------------
WARNING  root:test_demo.py:6 this is warning log...
ERROR    root:test_demo.py:7 this is error log...
CRITICAL root:test_demo.py:8 this is critical log...
==================== short test summary info ====================
FAILED test_demo.py::test_demo - assert 'ERROR' != 'ERROR'
==================== 1 failed in 0.08s ====================

(demo-HCIhXOHq) E:\demo>
```

11.7 Pytest 如何进行正确配置及使用日志功能

在 Pytest 自动化测试中，如果只简单从应用的角度来说，完全可以不去了解 Pytest 中显示的信息部分以及原理，完全可以使用 Pytest.ini 进行配置，从而完成比较通用的日志配置。

这里，我们推荐使用下面的配置，其中 log_cli 相关的四条配置是用来配置 live log 实时日志的，而其他三条配置则是用例配置 capture log，即捕获日志的。这里分别对他们的日志级别、日志格式、时间戳格式进行了设置，日志级别都设置为 info，当然如果脚本稳定之后，提交自动化测试代码仓库时，可以将日志级别调整为 warning。

```
[pytest]
log_cli = True
log_cli_level = info
log_cli_format = %(asctime)s | %(levelname)s | %(filename)s:%(lineno)s | %(message)s
log_cli_date_format = %Y-%m-%d %H:%M:%S

log_level = info
log_format = %(asctime)s | %(levelname)s | %(filename)s:%(lineno)s | %(message)s
log_date_format = %Y-%m-%d %H:%M:%S
```

下面用一个简单的测试脚本来展示一下上述日志配置的效果。

```
import logging

def test_demo():
    logging.debug("this is debug log ...")
    logging.info("this is info log ...")
    logging.warning("this is warning log ...")
    logging.error("this is error log ...")
    logging.critical("this is critical log ...")
    assert 1 == 2
```

可以看出，这里显示了实时日志（live log），未显示 debug 级别的日志，捕获日志（capture log）同样也未显示 debug 级别的日志，而且时间戳和日志格式相对来说都比较符合实际应用的，因此，这里推荐的 Pytest.ini 中对日志的配置，大家完全可以直接使用。

```
(demo-HCIhXOHq) E:\demo> pytest
==================== test session starts ====================
platform win32 -- Python 3.7.9, pytest-7.2.0, pluggy-1.0.0
```

```
rootdir: E:\demo, configfile: pytest.ini
plugins: assume-2.4.3, rerunfailures-10.2
collected 1 item

test_demo.py::test_demo
----------------------- live log call -----------------------
2022-12-06 00:42:06 | INFO | test_demo.py:5 | this is info log ..."
2022-12-06 00:42:06 | WARNING | test_demo.py:6 | this is warning log ..."
2022-12-06 00:42:06 | ERROR | test_demo.py:7 | this is error log ..."
2022-12-06 00:42:06 | CRITICAL | test_demo.py:8 | this is critical log ..."
FAILED                                                    [100%]

========================= FAILURES =========================
_____ test_demo _____

    def test_demo():
        logging.debug("this is debug log ...")
        logging.into("this is info log ...")
        logging.warning("this is warning log ...")
        logging.error("this is error log ...")
        logging.critical("this is critical log ...")
>       assert 1 == 2
E       assert 1 == 2

test_demo.py:9: AssertionError
--------------------- Captured log call ---------------------
2022-12-06 00:42:06 | INFO | test_demo.py:5 | this is info log ..."
2022-12-06 00:42:06 | WARNING | test_demo.py:6 | this is warning log ..."
2022-12-06 00:42:06 | ERROR | test_demo.py:7 | this is error log ..."
2022-12-06 00:42:06 | CRITICAL | test_demo.py:8 | this is critical log ..."
================= short test summary info =================
FAILED test_demo.py::test_demo - assert 1 == 2
================== 1 failed in 0.07s ==================

(demo-HCIhXOHq) E:\demo>
```

第 12 章　Allure 测试报告

Allure 是一个用于生成 Pytest 自动化测试报告的框架,且可以生成 web 页面。

12.1　Windows10 安装 allure

首先,检查系统是否已经正确安装 jdk,我们打开 cmd 窗口,然后执行 java-version,如图 12-1 所示,显示已经安装 jdk。

```
C:\Users\Administrator>java -version
java version "1.8.0_212"
Java(TM) SE Runtime Environment (build 1.8.0_212-b10)
Java HotSpot(TM) 64-Bit Server VM (build 25.212-b10, mixed mode)

C:\Users\Administrator>
```

图 12-1　执行 java-verison 显示结果

接着,继续在 cmd 窗口中执行 javac 命令,如图 12-2 所示,表示 jdk 已经正确安装了配置。

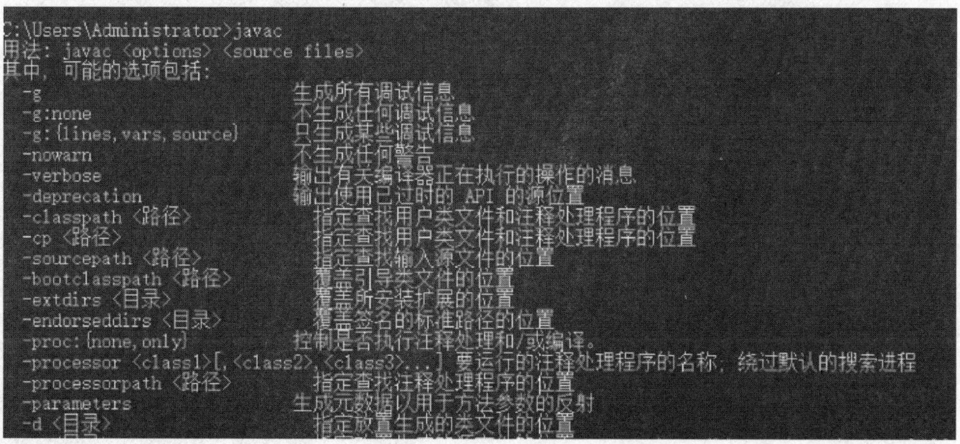

```
C:\Users\Administrator>javac
用法: javac <options> <source files>
其中, 可能的选项包括:
  -g                         生成所有调试信息
  -g:none                    不生成任何调试信息
  -g:{lines,vars,source}     只生成某些调试信息
  -nowarn                    不生成任何警告
  -verbose                   输出有关编译器正在执行的操作的消息
  -deprecation               输出使用已过时的 API 的源位置
  -classpath <路径>          指定查找用户类文件和注释处理程序的位置
  -cp <路径>                 指定查找用户类文件和注释处理程序的位置
  -sourcepath <路径>         指定查找输入源文件的位置
  -bootclasspath <路径>      覆盖引导类文件的位置
  -extdirs <目录>            覆盖所安装扩展的位置
  -endorseddirs <目录>       覆盖签名的标准路径的位置
  -proc:{none,only}          控制是否执行注释处理和/或编译。
  -processor <class1>[,<class2>,...]  要运行的注释处理程序的名称; 绕过默认的搜索进程
  -processorpath <路径>      指定查找注释处理程序的位置
  -parameters                生成元数据以用于方法参数的反射
  -d <目录>                  指定放置生成的类文件的位置
```

图 12-2　执行 javac 命令显示结果

然后,下载 Alure 的安装包,打开 github 上 allure 包的发布地址:https://github.com/allure-framework/allure2/releases,如图 12-3,单击下载 zip 包。

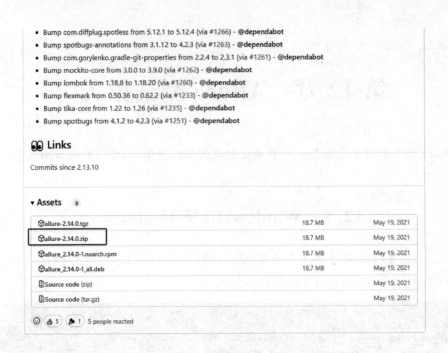

图 12-3 allure 下载安装包下载地址

之后，将下载好的 zip 包解压，并拷贝至安装位置，比如这里选择放在 D:\ProgramFile\ 目录下，如图 12-4 所示。

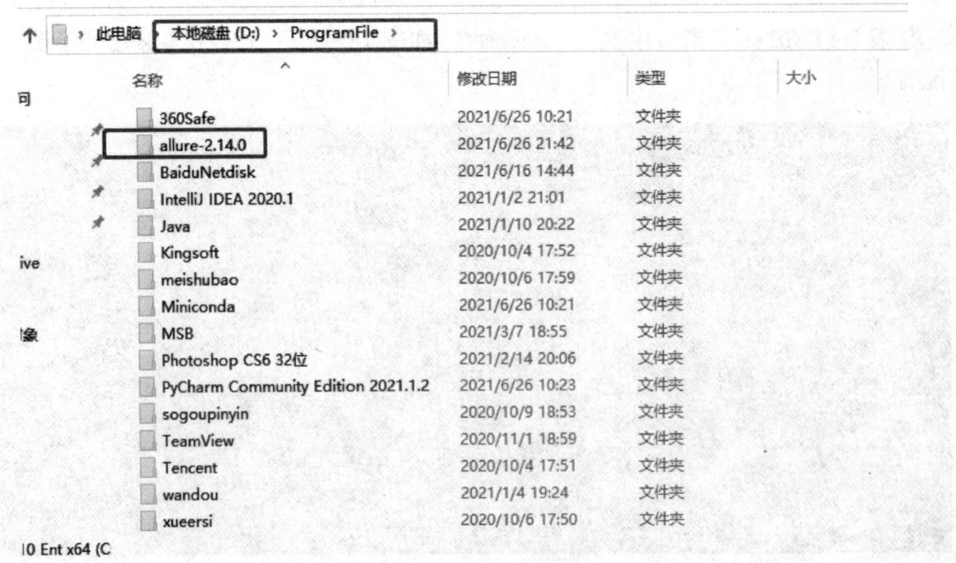

图 12-4 alure 解压目录

配置 Allure 的环境变量，如图 12-5 所示，将 D:\ProgramFile\allure-2.14.0\bin 添加到 Path 变量中去。

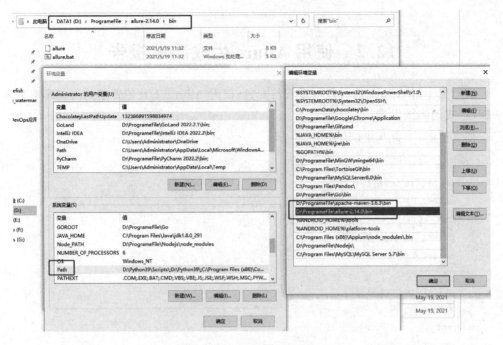

图 12-5 配置 allure 环境变量

双击 Allure.bat 文件,执行,如图 12-6 所示。

图 12-6 allure.bat 文件位置

打开 cmd 窗口,输入 Allure-version 如图 12-7 所示,已经安装成功。

```
C:\Users\Administrator>allure --version
2.14.0
C:\Users\Administrator>
```

图 12-7 查看 allure 版本号

通过下面命令安装 Allure-Pytest 插件。

```
pip install allure-pytest
```

12.2 使用 Allure 生成测试报告

这里先不对 Allure 进行任何设置与配置，仅用 Allure 工具使普通的测试脚本生成 WEB 形式的测试报告。

如下所示，准备 6 个测试脚本。

```
def test_1():
    print("in test_1")
    assert 1 == 1
def test_2():
    print("in test_2")
    assert 1 == 2
def test_3():
    print("in test_3")
    assert 1 == 1
def test_4():
    print("in test_4")
    assert 1 == 2
def test_5():
    print("in test_5")
    assert 1 == 1
def test_6():
    print("in test_4")
    assert 1 == 2
```

若在控制台执行脚本，则仅会得到如下执行结果。当然，对代码开发人员，没有问题，但是对于测试人员，或者从更加人性化角度来说，这样的输出是无法满足我们需求的。

```
(demo-HCIhX0Hq) E:\demo> pytest -s
==================== test session starts ====================
platform win32 -- Python 3.7.9, pytest-7.2.0, pluggy-1.0.0
benchmark: 4.0.0 (defaults: timer=time.perf_counter disable_gc=False min_rounds=5 min_time=0.000005 max_time=1.0 calibration_precision=10 warmup=False warmup_iterations=100000)
rootdir: E:\demo, configfile: pytest.ini
plugins: allure-pytest-2.12.0, assume-2.4.3, attrib-0.1.3, benchmark-4.0.0, rerunfailures-10.2
collected 6 items

test_demo.py in test_1
```

```
. in test_2
Fin test_3
. in test_4
Fin test_5
. in test_4
F


========================= FAILURES =========================
_____ test_2 _____

    def test_2():
        print("in test_2")
>       assert 1 == 2
E       assert 1 == 2

test_demo.py:6: AssertionError
_____ test_4 _____

    def test_4():
        print("in test_4")
>       assert 1 == 2
E       assert 1 == 2

test_demo.py:12: AssertionError
_____ test_6 _____

    def test_6():
        print("in test_4")
>       assert 1 == 2
E       assert 1 == 2

test_demo.py:18: AssertionError
================== short test summary info ==================
FAILED test_demo.py::test_2 - assert 1 == 2
FAILED test_demo.py::test_4 - assert 1 == 2
FAILED test_demo.py::test_6 - assert 1 == 2
================ 3 failed, 3 passed in 0.11s ================

(demo-HCIhXOHq) E:\demo>
```

此时，就可以借助 Allure 工具先生成自动化测试报告所需要的执行数据，执行如下命令。

```
pytest -s --alluredir=./temp
```

然后，在当前目录 temp 目录下生成采集数据文件，如图 12-8 所示。

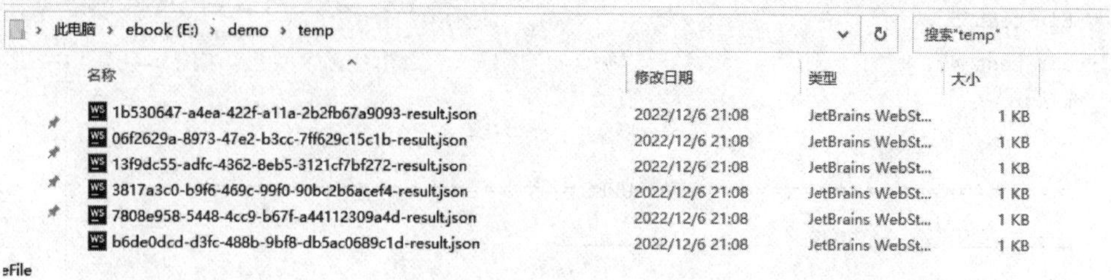

图 12-8 生成 allure 报告的数据文件

之后，继续执行如下命令，将生成报告的数据生成测试报告的 WEB 页面。

```
allure generate ./temp -o ./report
```

接着，生成测试报告 WEB 相关文件，存放在 report 目录下，report 目录如图 12-9 所示。

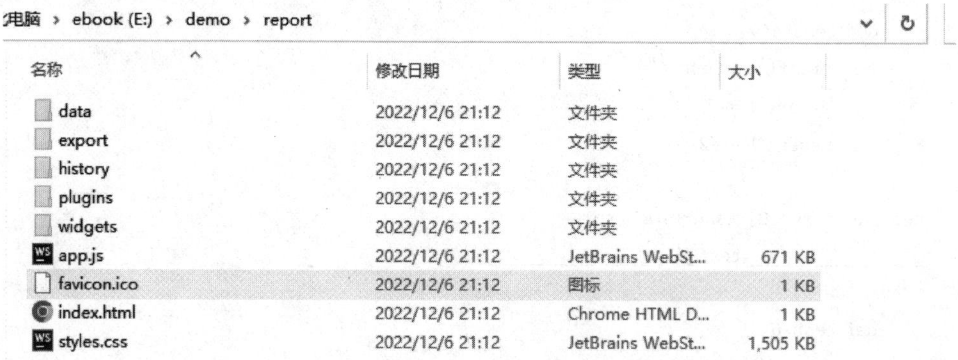

图 12-9 report 目录内容

我们发现在 report 目录中有 index.html 文件，由于 report 目录中的文件都是静态文件，为了能从浏览器中看到测试报告的内容，还需要执行如下命令启动一个内置的服务，进而在线查看测试报告。

```
allure open report
```

此时，在浏览器中打开 index.html，看到了 Allure 生成的自动化测试报告，如图 12-10 所示。

单击图 12-11 中深色框所在位置，设置中文页面显示。

如图 12-12 所示，单击【测试套】按钮，然后在测试套中单击一个用例，在右侧会显示出该用例断言失败的内容。

Allure 测试报告 第 12 章

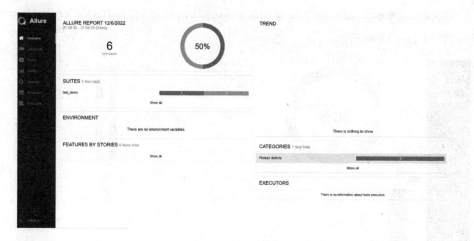

图 12-10　Allure 生成的报告

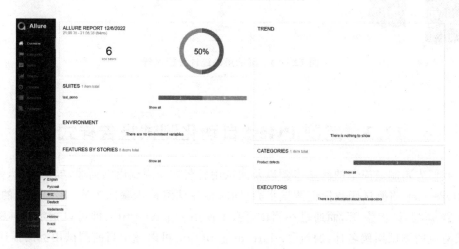

图 12-11　设置页面为中文

图 12-12　测试用例报告详细内容

单击【图表】按钮,则出现了测试用例的统计分析结果,并以环状图、条形图等形式显示,如图 12-13 所示。

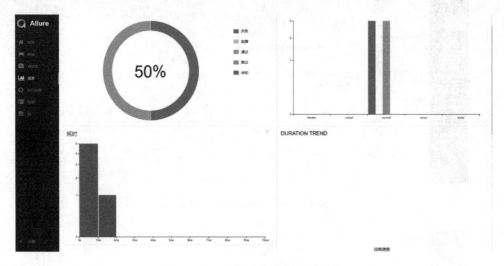

图 12-13　测试用例统计分析报告

12.3　定制 Pytest 自动化测试报告样式

我们先定制最简单的测试步骤以及测试用例名称和测试用例描述,如下所示,使用 @allure.step 将测试函数标记测试步骤,同时标注此测试步骤的内容,比如这里的"测试步骤 1""测试步骤 2",而通过在测试函数上使用 @allure.title 即可以指定当前测试函数对应的测试用例名称,通过 @allure.descrtiption 可以标记当前测试用例的描述。

```
import pytest
import allure
import logging


@allure.step("测试步骤 1")
def step_1():
    logging.info("测试步骤 1 中的功能实现 ...")


@allure.step("测试步骤 2")
def step_2():
    logging.info("测试步骤 2 中的功能实现 ...")


@allure.title("测试用例标题:我的第一个测试用例")
@allure.description("测试用例描述:我的第一个测试用例的描述,主要是为了展示测试脚本
```

如何使用 allure 生成漂亮的测试报告")
    def test_01():
        step_1()
        step_2()

通过执行有的命令,即可生成测试报告数据采集文件,且这些文件均为 json 文件,注意加上 - clean-alluredir,否则如果目录中已经存在数据,会在后面生成测试报告页面的时候把老数据也显示出来,进而带来不必要的干扰命令,如下所示。

```
pytest -- alluredir = ./temp -- clean-alluredir
```

生成数据采集文件后,再次执行如下的命令,可生成 WEB 页面的自动化测试报告,这里报告的相关文件均存在 report 目录下,注意要加上 -- clean 参数,否则也会将之前的老数据显示到页面上。

```
allure generate ./temp - o ./report -- clean
```

最后,通过如下命令即可打开 web 页面形式的自动化测试报告。

```
allure open report
```

打开 web 页面,测试报告的测试用例标题、描述、步骤名称如图 12 - 14 所示。

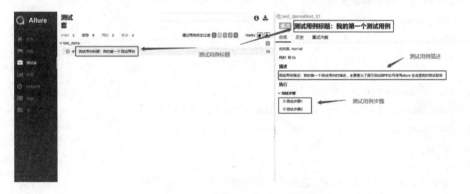

图 12 - 14　测试用例描述、标题、步骤

可以看到,此时的测试报告已经可以基本描述测试用例的具体内容了,相对于比较完整的测试报告而言,这里还缺少一些用例的属性,比如用例属于产品的什么功能,关联的用户故事又是什么,是否关联 bug,当前的用例等级又是什么等。下面我们继续丰富自动化测试报告的内容。

为了更好地显示测试报告的效果,我们又增加了一个测试用例,同时,在测试用例函数上方增加了用例级别、用户故事、链接、issue 及其功能。这里用例级别常用的有 blocker,critical,normal,minor,trivial 等级别,程度依次降低如下所示。

```
import pytest
import allure
```

```python
import logging


@allure.step("测试步骤 1")
def step_1():
    logging.info("测试步骤 1 中的功能实现...")


@allure.step("测试步骤 2")
def step_2():
    logging.info("测试步骤 2 中的功能实现...")


@allure.feature("allure 功能")
@allure.issue(url = "www.baidu.com", name = "需求来源于 xxx")
@allure.link(url = "www.baidu.com", name = "需求文档地址")
@allure.story("allure 应用场景")
@allure.severity('normal')
@allure.title("测试用例标题:我的第一个测试用例")
@allure.description("测试用例描述:我的第一个测试用例的描述,主要是为了展示测试脚本如何使用 allure 生成漂亮的测试报告")
def test_01():
    step_1()
    step_2()


@allure.feature("allure 功能")
@allure.issue(url = "www.baidu.com", name = "需求来源于 xxx")
@allure.link(url = "www.baidu.com", name = "需求文档地址")
@allure.story("allure 应用场景")
@allure.severity('critical')
@allure.title("测试用例标题:我的第二个测试用例")
@allure.description("测试用例描述:我的第二个测试用例的描述,主要是为了展示测试脚本如何使用 allure 生成漂亮的测试报告")
def test_02():
    step_1()
    step_2()
```

依次执行如下命令,重新生成测试报告。

```
pytest --alluredir=./temp --clean-alluredir
allure generate ./temp -o ./report --clean
allure open
```

然后,打开测试报告页面,切换到【功能】页面,即可看到 feature 功能、story 用户故事、链接、用例优先级、用例标题、用例描述、用例步骤等,如图 12-15 所示。

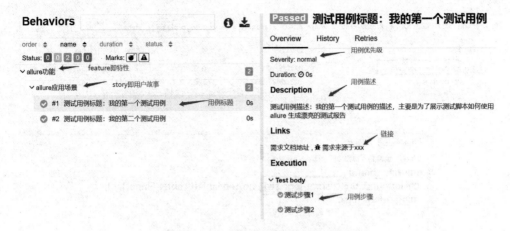

图 12-15 测试报告定制化详情

12.4 定制 Allure 报告中的 logo

在企业应用实践中,当对自动化测试做得相对比较成熟后,常会希望自动化测试报告能够嵌入自己公司的 logo 或者产品的相关图片,此时就要修改 Allure 报告中的 logo。

修改 Allure 报告中的 logo 很简单,首先将要使用的 logo 图片复制到 allure 安装的目录下,如图 12-16 所示。

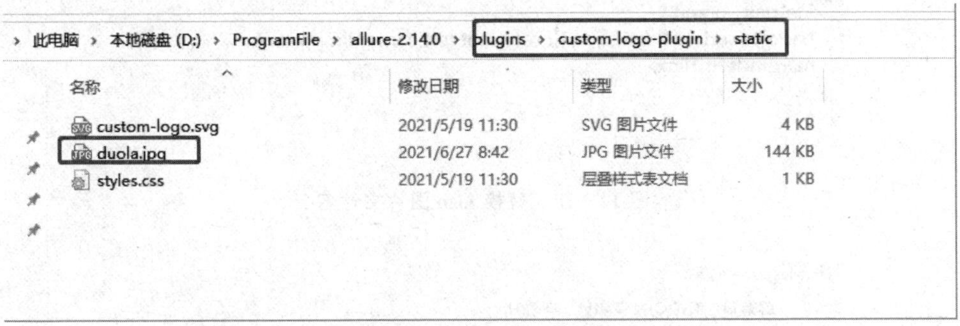

图 12-16 logo 图片替换位置

然后,打开 styles.css 文件,找到当前 logo 的文件名,如图 12-17 所示。
再将 custom-logo.svg 替换为自己的 logo 图片名称,如图 12-18 所示。
之后,调整一下尺寸,如图 12-19 所示。
打开加载插件的配置文件,默认内容如图 12-20 所示。

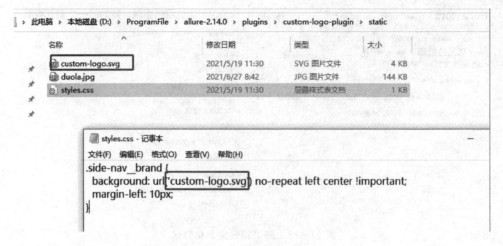

图 12-17 默认 logo 文件名

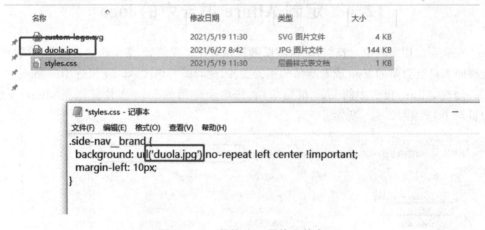

图 12-18 替换 logo 图片文件名

图 12-19 调整尺寸

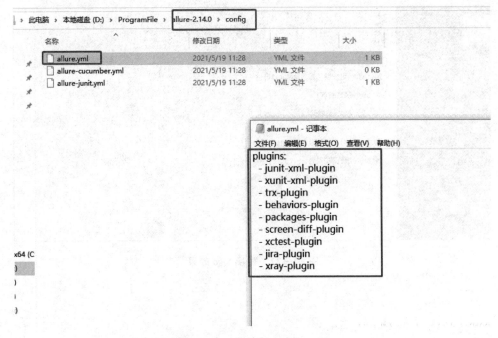

图 12-20　加载插件配置文件

接着，将替换 logo 的插件名称填写进来，如图 12-21 所示。

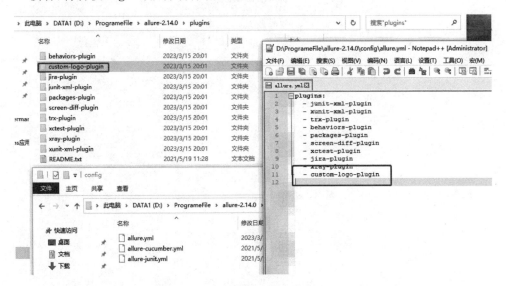

图 12-21　增加替换 logo 的插件

再次刷新页面，即可看到此时自动化测试报告的 logo 已经变了，如这里替换为了一个哆啦 A 梦的图像，如图 12-22 所示。

图 12-22 替换 logo 效果

第 13 章 与 jenkins 持续集成

至此，Pytest 自动化框架的知识点就介绍的差不多了，本章就基于 Pytest 的基础知识点，教大家一步一步地设计真实的自动化测试框架，并将脚本上传至 github，同时安装部署 jenkins，且在 jenkins 上配置自动从 github 拉取代码，执行测试脚本，同时生成 Allure 测试报告。

13.1 设计开发自动化测试框架

这里就一步一步设计一下实用的自动化测试框架。我们将要设计的测试框架命名为 eTest，首先创建 eTest 目录。为了考虑后续扩展性，比如后期如果觉得测试框架可以发布为 Python 包，此时就要考虑好，因此我们在 eTest 目录中继续创建一个 eTest 目录，当然暂时先不考虑发布包的方式，但是要为后续可能的功能增强做好兼容性扩展性设计。后面的操作将都在 eTest/eTest/目录下执行。

首先在 eTest/eTest/目录下创建 tests 文件夹，用于存放自动化测试脚本，然后再创建 libs 文件夹，用于存放公共的函数封装，libs 目录下这里就不再进行深入讲解了，主要根据具体业务的自动化脚本做好恰当的函数库封装，公共的框架，这里就不再进一步深入介绍了。

至此，需要说明一点，后续自动化脚本的执行位置应定位在 eTest/eTest/tests 目录下，即 eTest/eTest/tests/目录相当于测试脚本的根目录，因此首先在 eTest/eTest/tests 目录下创建一个 Pytest.ini 文件，用于配置通用的公共的自动化脚本。

我们来配置测试函数、测试类、测试文件的命名规则。虽然一般来说不推荐对默认的命名规则进行修改，但由于默认的规则中存在一个很容易混淆的点，即 Pytest 默认要求测试文件必须是 test_*.py 或者 *_test.py 格式，测试方法必须是 test 开头，测试必须是 Test 开头。这里面就涉及了一些问题，比如有的以 test_开头要求带下画线，有的则没有要求下画线，有的可以以_test 结尾，有的又不能，这就很容易引起混乱，因此这里完全可以把范围扩大一点，即将测试用例文件的规则修改为以 test_*.py 或 *_test.py，测试函数以 test_开头或者以_test 结尾，测试类以 Test 开头或 Test 结尾，这样就很容易记忆了，而且也能包括默认的 Pytest 规则。是否带下划线完全根据 Python 代码规范要求，即文件名或者测试函数名均为小写，必须要带下划线，而测试类首字母大写，就来定不需要带下画线，这样既能包含默认的 Pytest 命名规则，又能做到风格统一，因此需要首先在 Pytest.ini 中进行如下配置：

```
[pytest]
python_files = test_*  *_test.py
python_classes =   Test*   *Test
python_functions = test_*   *_test
```

接下来，注册测试脚本中使用的标签，因为如果不注册标签，在执行的时候会产生大量的告警，因此这里事先注册几个常用通用的标签，标签应按照通常的测试分类进行注册，即在 Pytest.ini 中增加 markers 的内容，增加后内容如下所示。

```
[pytest]
python_files = test_*  *_test.py
python_classes =   Test*   *Test
python_functions = test_*   *_test

markers =
    api: api tests
    smoke: smoke tests
    function: function tests
    system: system tests
    security: security tests
    performance: performance tests
```

接下来配置日志，这里将实时日志和捕获日志全部配置为推荐配置，如下所示。

```
[pytest]
python_files = test_*  *_test.py
python_classes =   Test*   *Test
python_functions = test_*   *_test

markers =
    api: api tests
    smoke: smoke tests
    function: function tests
    system: system tests
    security: security tests
    performance: performance tests

log_cli = True
log_cli_level = info
log_cli_format = %(asctime)s | %(levelname)s | %(filename)s:%(lineno)s | %(message)s"
log_cli_date_format = %Y-%m-%d %H:%M:%S
log_level = info
log_file = eTest.log
log_format = %(asctime)s | %(levelname)s | %(filename)s:%(lineno)s | %(message)s"
```

```
log_date_format = %Y-%m-%d %H:%M:%S
```

接下来,配置 Pytest 命令默认的参数,我们将-s 参数加到默认的参数中,用户如果想增加配置,可以直接在 adopts 后面增加即可。

```
[pytest]
python_files = test_*  *_test.py
python_classes =  Test*  *Test
python_functions = test_*  *_test

markers =
    api: api tests
    smoke: smoke tests
    function: function tests
    system: system tests
    security: security tests
    performance: performance tests

log_cli = True
log_cli_level = info
log_cli_format = %(asctime)s | %(levelname)s | %(filename)s:%(lineno)s | %(message)s
log_cli_date_format = %Y-%m-%d %H:%M:%S
log_level = info
log_file = eTest.log
log_format = %(asctime)s | %(levelname)s | %(filename)s:%(lineno)s | %(message)s
log_date_format = %Y-%m-%d %H:%M:%S

addopts =-s
```

至此,Pytest.ini 的配置基本就能满足通常情况下的场景了。

然后,在 eTest/eTest/tests/目录下创建 env.py 文件,用于存放测试环境信息,目的在于使测试脚本与环境解耦,即通过修改 env.py 中的环境信息使脚本迁移环境,这一点对于设计自动化测试框架是至关重要的。此外,因为 Python 文件在引入的时候就会自动执行文件,因此建议使用类来组织变量,即在 eTest/eTest/tests/env.py 中创建 Env 类,在类中定义变量,比如定义 ip、username、password。

```
class Env:
    ip = "127.0.0.1"
    username = "root"
    password = "root"
```

接下来,在 eTest/eTest/tests/目录下创建 conftest.py 文件,在 conftest.py 中定义一个用于传递环境变量值的 fixture 即 env,如下所示。

```python
import pytest
from .env import Env


@pytest.fixture(scope="session")
def env():
    return Env
```

然后,在 conftest.py 中写一个 session 级别的 fixture,以用于编写 session 级别的 setup 和 teardown,此时 conftest.py 的代码如下所示。

```python
import pytest
import logging
from .env import Env


@pytest.fixture(scope="session")
def env():
    return Env


@pytest.fixture(scope="session", autouse=True)
def session_setup_teardown():
    logging.info("开始执行 session 级别的 setup ...")
    # session 级的 setup,可在下面继续编写 session 级别的 setup 操作

    yield
    logging.info("开始执行 session 级别的 teardown ...")
    # session 级别的 teardown,可在下面继续编写 session 级别的 teardown 的操作
```

接着,在 eTest/eTest/tests/目录下依次创建目录 api-test、function-test、performance-test、security-test、system-test,依次存放 api 测试用例、功能测试用例、性能测试用例、安全测试用例、系统测试用例。注意,这里创建的目录中都默认创建一个__init__.py 文件,标志着这些目录均为可导入的包。在 eTest/eTest/tests/api-test/目录下继续创建一个脚本模板,如 test_api_example.py,并编写脚本模板代码,这里使用 Allure 生成报告,而使用 logging 打印日志。之后续继续编写脚本,直接新建脚本文件,将模板拷贝过去,然后进行修改填充,此外,这里还演示了如何使用全局环境变量。

```python
import pytest
import allure
import logging


@allure.step("测试步骤 1:xxx")
def step_1(env):
    logging.info("测试步骤 1:xxx ...")
```

```python
    # 编写步骤1的内容,比如这里打印 env.ip
    logging.info(env.ip)

@allure.step("测试步骤2:xxx")
def step_2(env):
    logging.info("测试步骤2:xxx ...")
    # 编写步骤1的内容,比如这里打印 env.username 和 env.password
    logging.info(env.username)
    logging.info(env.password)


@allure.feature("功能:xxx")
@allure.issue(url = "www.xxx.com",name = "需求来源于xxx")
@allure.link(url = "www.xxx.com",name = "需求文档地址")
@allure.story("用户故事:xxx")
@allure.severity('normal')
@allure.title("测试用例:xxxc")
@allure.description("测试用例描述:xxx")
def test_api_01(env):
    step_1(env)
    step_2(env)
```

最后,打开 cmd 窗口,进入 eTest 目录下,通过如下命令创建一个基于 Python3.7 版本的虚拟机环境。

```
pipenv – python 3.7
```

执行如下命令即可激活此虚拟环境。

```
pipenv shell
```

安装依赖。

```
pip install pytest
pip install allure-pytest
```

执行如下命令,生成依赖文件。

```
pip freeze > requirements.txt
```

此时,在 eTest/目录下生成 requirements.txt 文件,其内容如下所示。

```
allure-pytest == 2.12.0
allure-python-commons == 2.12.0
attrs == 22.1.0
colorama == 0.4.6
exceptiongroup == 1.0.4
```

```
importlib-metadata==5.1.0
iniconfig==1.1.1
packaging==22.0
pluggy==1.0.0
pytest==7.2.0
six==1.16.0
tomli==2.0.1
typing_extensions==4.4.0
zipp==3.11.0
```

然后,在cmd中切换到eTest/eTest/tests目录下,执行Pytest命令,结果如下所示,可见所有日志配置均已生效。

```
(eTest-OXFOQ-1z) G:\github\eTest\eTest\tests> pytest
==================== test session starts ====================
platform win32 -- Python 3.7.9, pytest-7.2.0, pluggy-1.0.0
rootdir: G:\github\eTest\eTest\tests, configfile: pytest.ini
plugins: allure-pytest-2.12.0
collected 1 item


api-test/test_api_example.py::test_api_01
-------------------- live log setup --------------------
2022-12-11 11:31:46 | INFO | conftest.py:11 | 开始执行session级别的setup..."
-------------------- live log call --------------------
2022-12-11 11:31:46 | INFO | test_api_example.py:8 | 测试步骤1:xxx..."
2022-12-11 11:31:46 | INFO | test_api_example.py:10 | 127.0.0.1"
2022-12-11 11:31:46 | INFO | test_api_example.py:14 | 测试步骤2:xxx..."
2022-12-11 11:31:46 | INFO | test_api_example.py:16 | root"
2022-12-11 11:31:46 | INFO | test_api_example.py:17 | root"
PASSED
-------------------- live log teardown --------------------
2022-12-11 11:31:46 | INFO | conftest.py:15 | 开始执行session级别的teardown..."


==================== 1 passed in 0.03s ====================

(eTest-OXFOQ-1z) G:\github\eTest\eTest\tests>
```

至此,一个简单使用基于Pytest自动化测试的框架就搭建完成了,如果想编写自动化脚本,只需要在eTest/eTest/tests目录下对应的测试类型的目录下进行脚本开发即可。

13.2　测试脚本上传 git 代码仓库

在企业实战中,自动化测试脚本也要放在代码管理平台,当然也可以选择第三方公共的 git 代码托管平台,比如 github、gitee 等。另外,也可以在企业内部搭建 gitlab 作为代码托管平台,使用方式都是类似的。这里为了后期继续同步优化测试框架,同时考虑以学习集成 Pytest 自动化框架为主,因此就以上传 github 为例进行讲解。

首先在 github 平台注册一个自己的账号,然后打开 cmd 窗口,执行如下命令,其中 xxx@163.com 为自己的邮箱。然后连续按回车即可。

```
ssh-keygen -t rsa -C "xxx@163.com"
```

此时,会在用户目录下的.ssh 目录中生成私钥和公钥,如使用 administrator 用户,则会在 C:\Users\Administrator\.ssh 目录下生成 id_rsa 文件和 id_rsa.pub 文件,其中 id_rsa 文件内容为私钥,id_rsa.pub 文件内容为公钥。

然后,登录 github,单击头像,在下拉表中单击"settings",如图 13-1 所示。

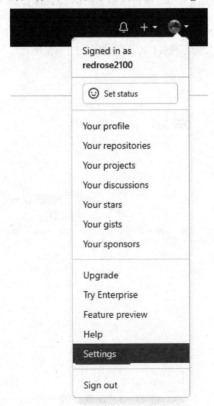

图 13-1　登录 github 后单击 settings

然后,单击左侧的"SSH and GPG keys",在右侧单击"New SSH key",如图 13-2 所示。

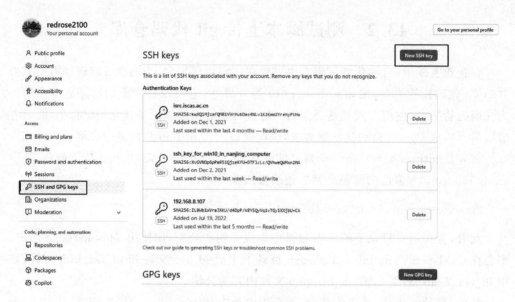

图 13-2 单击 SSH and GPG keys 页面

设置标题,标题可以自由命名,将生成的公钥即 id_rsa.pub 文件的内容全部拷贝至 Key 的输入框中,单击"Add SSH Key",如图 13-3 所示。

图 13-3 配置公钥

然后,创建代码仓库,单击右上角的"+",再单击"New Repository",如图 13-4 所示,开始创建仓库。

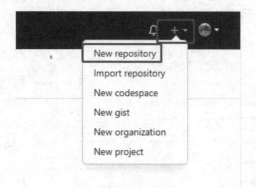

图 13-4　github 创建仓库

然后,配置仓库,如图 13-5 所示,设置仓库名,填写仓库描述,根据需求设置仓库公开或者私有,这里因为测试框架后续有可能继续优化,同时考虑便于查看,选择设置公开,然后在 Add .gitignore 选项中选择 Python 语言,如为开源协议就选择 MIT 协议,然后,单击"Create repository"创建仓库。

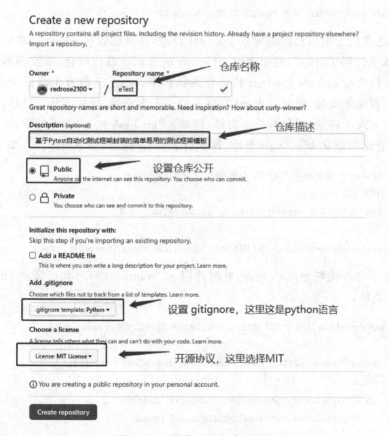

图 13-5　配置 github 仓库信息

如图 13-6 所示，打开代码仓，单击"Code"，再选择"SSH"，之后单击右边的复制按钮，即可复制基于 ssh 协议的链接了。

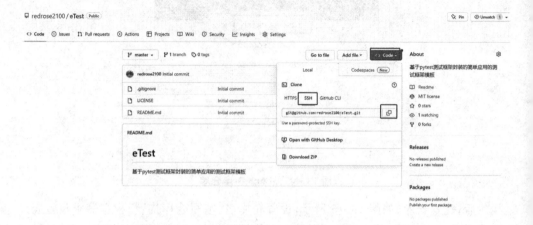

图 13-6　复制代码仓链接

打开 git bash 窗口，通过如下命令下载代码仓代码。

git clone git@github.com:redrose2100/eTest.git

当然，这里由于前面已经创建好了 eTest 目录，并且已经在 eTest 目录中设计好测试框架的基础代码了，因此，这里通过 git clone 的方式下载代码仓，需要将本地已经存在的 eTest 目录与 github 上 eTest 代码仓关联起来。当然这里有个前提，已经安装好git 客户端，若尚未安装，则可以去 git 官网下载一个客户端，然后安装。然后，在资源浏览器中打开 eTest 目录，用鼠标单击右键，再单击"git bash here"即可打开 git 命令行窗口，将当前仓库初始化为一个 git 仓库，即执行 git init 命令，如下所示。

```
Administrator@redrose2100-NJ MINGW64 /g/github/eTest
$ git init
Initialized empty Git repository in G:/github/eTest/.git/
```

```
Administrator@redrose2100-NJ MINGW64 /g/github/eTest (master)
```

通过如下命令关联 github 仓库，此时使用 git remote -v 可看到已经和 github 上的eTest 仓库关联。

```
$ git remote add origin git@github.com:redrose2100/eTest.git
```

```
Administrator@redrose2100-NJ MINGW64 /g/github/eTest (master)
$ git remote -v
origin  git@github.com:redrose2100/eTest.git (fetch)
origin  git@github.com:redrose2100/eTest.git (push)
```

```
Administrator@redrose2100-NJ MINGW64 /g/github/eTest (master)
```

因为github仓库的eTest仓库中在创建的时候设置了.gitnore及README.md文件,即github仓库上已经存在内容了,因此需要首先将github上的内容pull到本地,即执行如下命令。

```
$ git pull origin master
remote: Enumerating objects: 5, done.
remote: Counting objects: 100% (5/5), done.
remote: Compressing objects: 100% (5/5), done.
remote: Total 5 (delta 0), reused 0 (delta 0), pack-reused 0
Unpacking objects: 100% (5/5), 2.26 KiB | 16.00 KiB/s, done.
From github.com:redrose2100/eTest
 * branch            master     -> FETCH_HEAD
 * [new branch]      master     -> origin/master

Administrator@redrose2100-NJ MINGW64 /g/github/eTest (master)
```

此时,本地可以看到,已经将github仓库上的内容更新了,如图13-7所示。

图13-7 github更新到本地后目录

然后,在git命令行中执行如下几条命令可将本地已经设计好的测试框架内容上传到github代码仓。

```
git add .
git commit -m "add eTest"
git push origin master
```

然后在github仓库即可看到已经同步了,如图13-8所示。

图 13-8 测试框架已经在 github 仓库同步

13.3 使用 docker 搭建 jenkins

首先,需要有台虚拟机,这里以 CentOS7 的虚拟机为例,然后通过 docker -- help 命令查看是否安装 docker,如果尚未安装 docker,则按照接下来的操作安装,如果已经安装了 docker 了,则直接跳过 docker 的安装步骤。

执行如下命令,卸载旧版本的 docker。

```
sudo yum remove docker \
                docker-client \
                docker-client-latest \
                docker-common \
                docker-latest \
                docker-latest-logrotate \
                docker-logrotate \
                docker-engine
```

执行如下命令安装基础依赖。

```
sudo yum install -y yum-utils
```

接下来,配置 docker 的 repo 源,执行如下命令即可。

```
sudo yum-config-manager \
    --add-repo \
    https://download.docker.com/linux/centos/docker-ce.repo
```

执行如下命令,安装 docker。

```
sudo yum install docker-ce docker-ce-cli containerd.io docker-compose-plugin
```

安装完成后,执行如下命令,启动 docker 服务,同时将 docker 设置为开机自启动。

```
sudo systemctl start docker
sudo systenctl enable docker
```

然后,执行 sudo docker run hello-world 命令,检查 docker 是否安装成功,若能有如下显示,表示 docker 已经安装成功了。

```
[root@osssc-dev-01 ~]# sudo docker run hello-world
Unable to find image 'hello-world:latest' locally
latest: Pulling from library/hello-world
2db29710123e: Pull complete
Digest: sha256:2498fce14358aa50ead0cc6c19990fc6ff866ce72aeb5546e1d59caac3d0d60f
Status: Downloaded newer image for hello-world:latest
Hello from Docker!
This message shows that your installation appears to be working correctly.
To generate this message, Docker took the following steps:
 1. The Docker client contacted the Docker daemon.
 2. The Docker daemon pulled the "hello-world" image from the Docker Hub.
    (amd64)
 3. The Docker daemon created a new container from that image which runs the
    executable that produces the output you are currently reading.
 4. The Docker daemon streamed that output to the Docker client, which sent it
    to your terminal.
To try something more ambitious, you can run an Ubuntu container with:
 $ docker run -it ubuntu bash
Share images, automate workflows, and more with a free Docker ID:
 https://hub.docker.com/
For more examples and ideas, visit:
 https://docs.docker.com/get-started/
[root@osssc-dev-01 ~]#
```

接下来,使用 docker 安装部署 jenkins,要先下载 jenkins 的镜像,执行命令如下所示。

```
docker pull jenkins/jenkins:lts
```

然后,创建 jenkins 挂载目录,执行如下命令。

```
mkdir -p /docker/jenkins/var/jenkins_home
chmod 777 /docker/jenkins/var/jenkins_home
```

使用 docker 启动 jenkins 的容器,我们可以自定义容器端口映射。

```
docker run -d -p 10002:8080 -p 10003:50000 -v /docker/jenkins/var/jenkins_home:/
```

var/jenkins_home -v /etc/localtime:/etc/localtime - -name jenkins jenkins/jenkins:lts

配置镜像加速器,如下所示。

cd /docker/jenkins/var/jenkins_home/
vi hudson.model.UpdateCenter.xml

将 url 修改为 https://mirrors.tuna.tsinghua.edu.cn/jenkins/updates/update-center.json,内容如下所示。

```
<?xml version='1.1' encoding='UTF-8'?>
<sites>
    <site>
        <id>default</id>
        <url>https://mirrors.tuna.tsinghua.edu.cn/jenkins/updates/update-center.json</url>
    </site>
</sites>
```

然后,在浏览器中打开 ip:10002,这里因为前面启动容器的时候设置了10002端口映射8080端口,因此,需要使用10002端口,这个大家可以自行定义。之后打开,如图 13-9 所示。

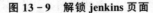

图 13-9　解锁 jenkins 页面

根据提示查看管理员密码,这里因为在启动容器的时候配置了目录挂载,即将/var/jenkins_home 目录挂载到本地的/docker/jenkins/var/jenkins_home:目录下,因此,这里直接使用如下命令即可查看管理员密码。

cat /docker/jenkins/var/jenkins_home/secrets/initialAdminPassword

输入管理员密码后,出现如图 13-10 所示页面,这里可以根据自己的具体情况选择。如果对 jenkins 已经比较熟悉了或者想一步一步学习如何安装插件,以及安装哪些插件等可以选择"选择插件来安装",如果为了快速完成工作或者担心出现各种问题可以选择"安装推荐的插件",这里暂时选择"安装推荐的插件"来说明,这样很多常用的插件就安装好了。

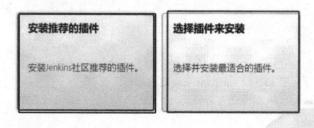

图 13-10 选择安装插件的方式

然后,设置管理员账号和密码,如图 13-11 所示。

图 13-11 设置管理员账号和密码

完成安装,如图 13-12 所示。

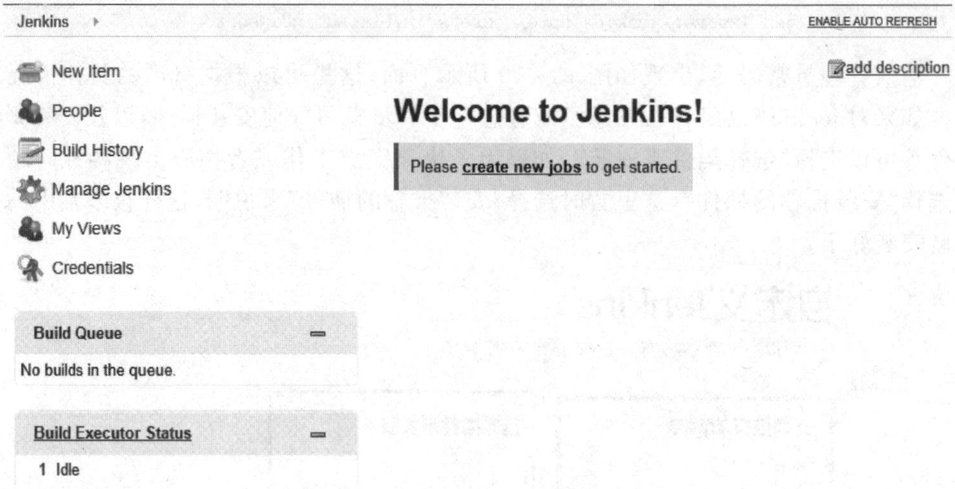

图 13-12　jenkins 安装完成后的页面

13.4　在 Centos7 系统中安装 git 客户端

因为需要下载测试脚本框架,因此我们的 centos 虚拟机上需要安装 git 工具,首先执行如下命令,安装依赖包。

```
yum install -y curl-devel expat-devel gettext-devel openssl-devel zlib-devel gcc-c++ perl-ExtUtils-MakeMaker
```

下载 git 安装包,如下所示。

```
wget https://mirrors.edge.kernel.org/pub/software/scm/git/git-2.31.1.tar.gz --no-check-certificate
```

下载完成后将 git 解压到 /usr/local/src 目录下,执行如下命令。

```
tar -zxvf git-2.31.1.tar.gz -C /usr/local/src/
```

然后,执行如下命令,编译,并安装 git。

```
cd /usr/local/src/git-2.31.1/
./configure --prefix=/usr/local/git
make
make install
```

执行如下命令,配置 git 环境变量。

```
echo "export PATH=$PATH:/usr/local/git/bin" >> /etc/profile
```

```
source /etc/profile
```

验证版本号是否正常。

```
[root@honghua-centos-02 ~]# git --version
git version 2.31.1
[root@honghua-centos-02 ~]#
```

需要注意,如果 centos 系统首次安装 git,通常会显示版本不对,这因为系统自带了一个低版本的 git。

```
[root@iZbp1flzt6x7pxmxfhmxeeZ git-2.31.1]# git --version
git version 1.8.3.1
[root@iZbp1flzt6x7pxmxfhmxeeZ git-2.31.1]#
```

此时,需要先卸载系统自带的 git,执行如下命令即可完成。

```
yum remove git -y
```

然后,再执行以下如下命令,使环境变量重新生效。

```
source /etc/profile
```

然后再次查看版本号,此时已经安装成功了。

```
[root@centos-02 ~]# git --version
git version 2.31.1
[root@centos-02 ~]#
```

此外,同样需要在 centos 虚拟机上生成公钥和私钥,然后查看是否已经存在公钥和私钥,如下所示,表示当前虚拟机已经存在公钥和私钥了。

```
[root@redrose2100 .ssh]# ls /root/.ssh/
id_rsa  id_rsa.pub  known_hosts
[root@redrose2100 .ssh]#
```

如果不存在 id_rsa 和 id_rsa.pub 文件或类似形式的其他加密方式的公钥和私钥文件,则可执行如下命令,注意这里面应将邮箱修改为自己的邮箱,连续按回车即可。

```
ssh-keygen -t rsa -C "xxx@163.com"
```

最后,就可以参考 13.2 节将公钥 id_rsa.pub 文件在 github 上配置公钥即可。

13.5　在 Centos7 系统中安装配置 Allure

首先,打开 Allure 的 github 地址,https://github.com/allure-framework/allure2/releases,然后找到 tgz 的发布包,单击右键,拷贝链接,如图 13-13 所示。

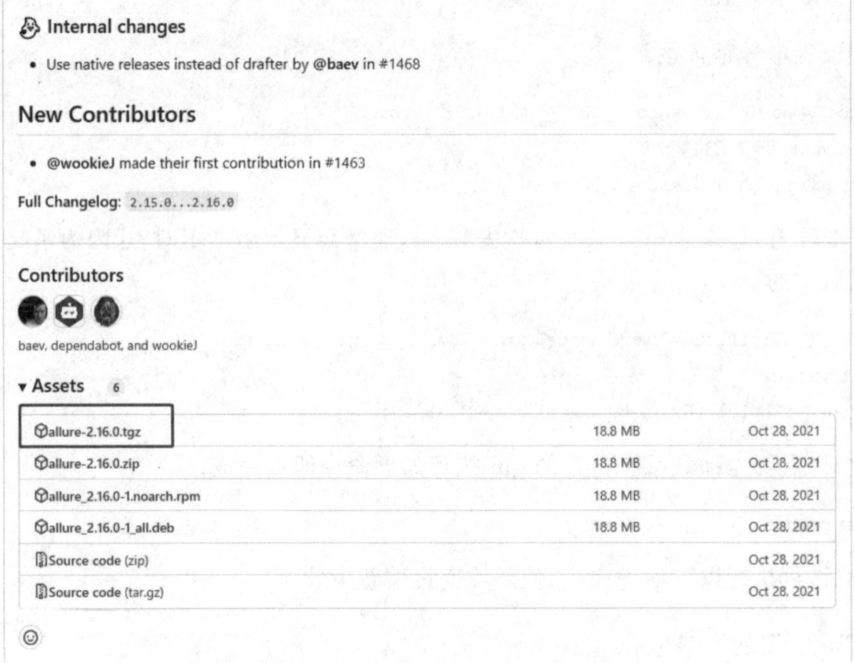

图 13-13　从 github 拷贝 allure 发布包的链接

在 CentOS7 虚拟机上使用 wget 下载包,如这里执行如下命令。

wget https://github.com/allure-framework/allure2/releases/download/2.16.1/allure-2.16.1.tgz

将 allure 解压到 /usr/local/ 目录下,即执行如下命令。

tar -zxvf allure-2.16.1.tgz -C /usr/local/

然后,编辑/etc/profile 文件,在文件末尾增加如下内容。

export PATH=$PATH:/usr/local/allure-2.16.1/bin

执行如下命令,使环境变量生效。

source /etc/profile

最后,验证 Allure 是否安装成功,如下所示表示安装成功。

```
[root@iZbp1flzt6x7pxmxfhmxeeZ bin]# allure --version
2.16.1
[root@iZbp1flzt6x7pxmxfhmxeeZ bin]#
```

13.6 Jenkins 基础配置

13.6.1 为 Jenkins 增加节点

首先，需要给 Jenkins 增加一个执行节点，在【Manage jenkins】→【Manage Plugins】→【Available】中搜索 SSH Agent 和 SSH Build Agents plugin 插件，安装成功后重启 jenkins，如图 13-14。

图 13-14　jenkins 安装 SSH Agent 插件

接着，配置 SSH Server，依次单击【系统管理】→【全局安全配置】，把 SSH Server 设置为启用（默认是禁用），如图 13-15 所示。

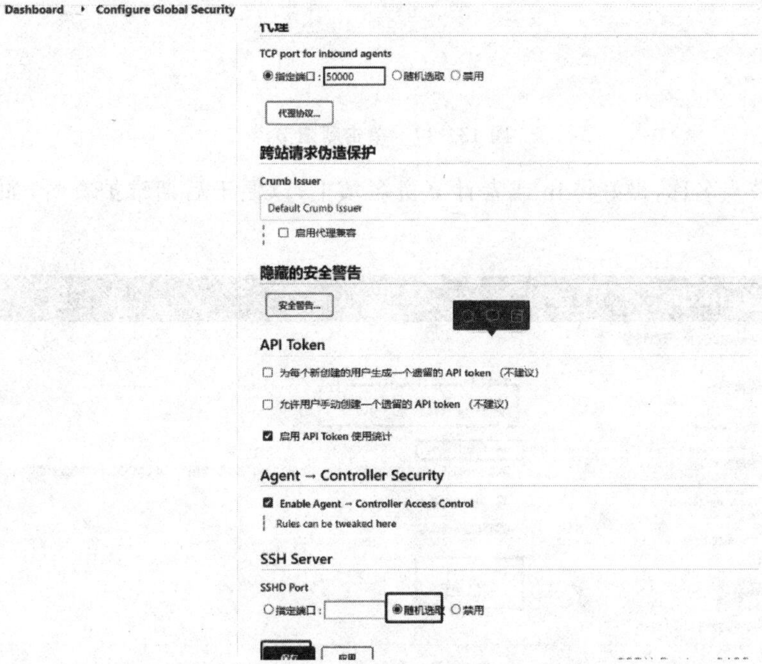

图 13-15　设置启动 SSH Server 配置

然后，回到 Jenkins 首页，单击【系统管理】→【节点管理】按钮，如图 13-16 所示。

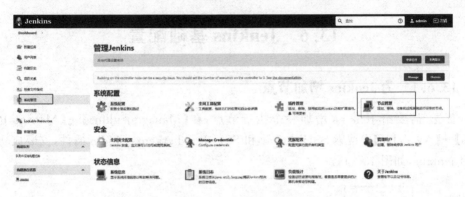

图 13-16　单击节点管理

单击【新建节点】，如图 13-17 所示。

图 13-17　单击新建节点

填写节点名称，最好将 ip 或者计算机名填上，以便于后期维护查看，如图 13-18 所示。

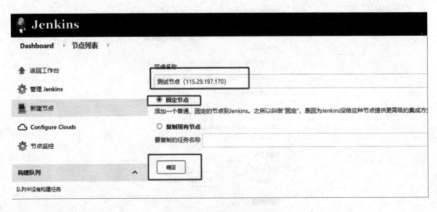

图 13-18　填写节点信息

之后，设置节点信息，如设置工作目录，可以设置为 /var/Jenkins/workspace，选择"launch agent via SSH"方式，再设置 ssh 登录节点的用户名和密码，在 Hos_Key 的确认策略选项选择"Non Verifying Verification Strategy"，如图 13-19 所示。然后，单击保存。

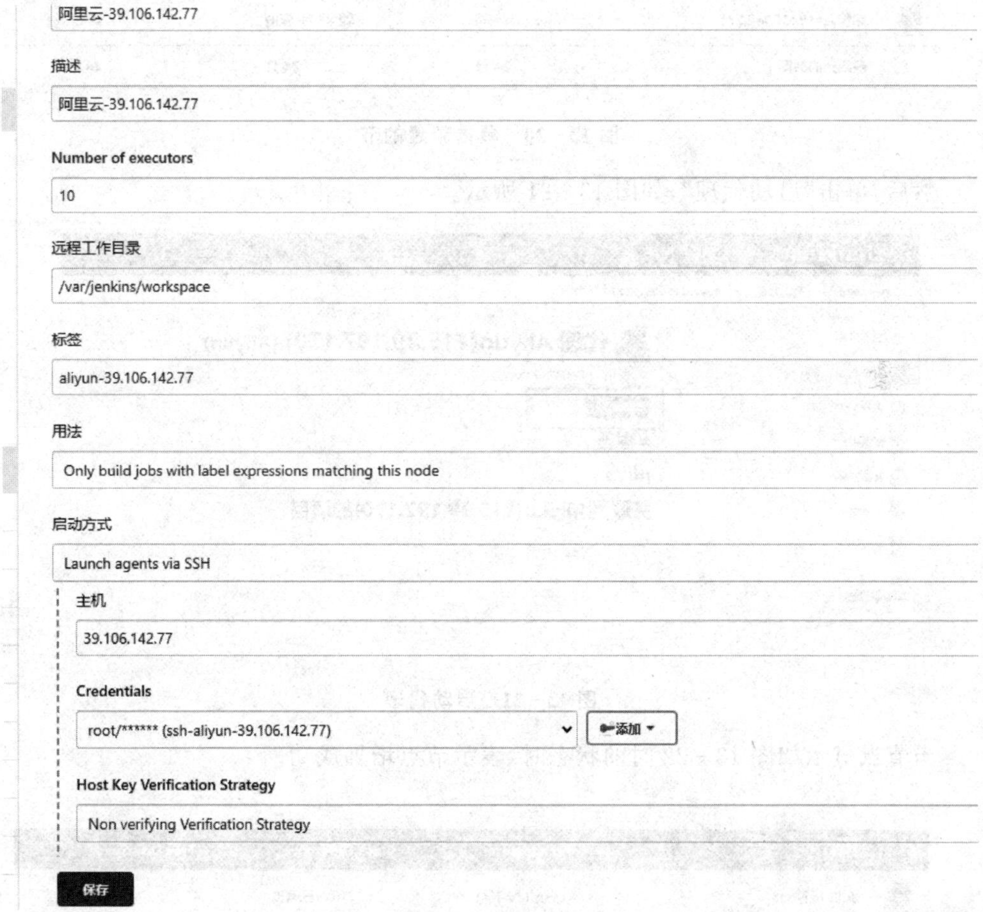

图 13-19　设置节点信息

在 CentOS 虚拟机上为 jenkins 设置 java 的链接，因为 jenkins 并没有使用 root 用户，而是在自己的工作空间下使用 jenkins 用户，因此需要给 jenkins 用户设置可用的 java 的链接。

```
# 这里 /var/jenkins/workspace 为上述步骤设置的节点的工作目录
mkdir -p /var/jenkins/workspace/jdk/bin/
which java
# which java 命令的结果,/usr/local/jdk1.8.0_301/bin/java,然后创建软连接
ln -s /usr/local/jdk1.8.0_301/bin/java /var/jenkins/workspace/jdk/bin/java
```

单击刚创建的节点，如图 13-20 所示。

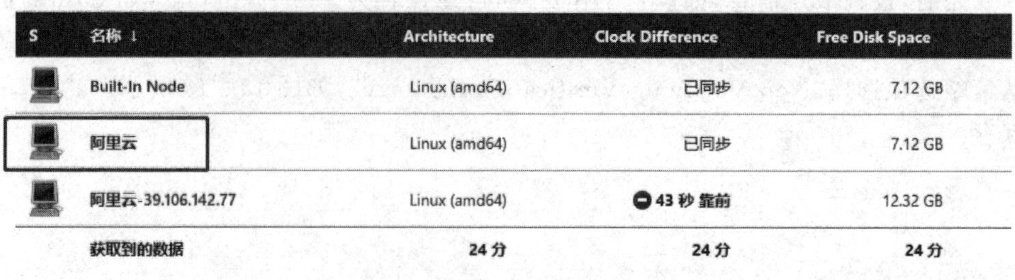

图 13-20 单击新建的节

然后,单击"启动代理",如图 13-21 所示。

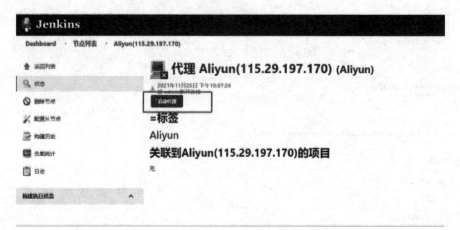

图 13-21 启动代理

当节点显示如图 13-22 时的状态时,表示节点增加成功。

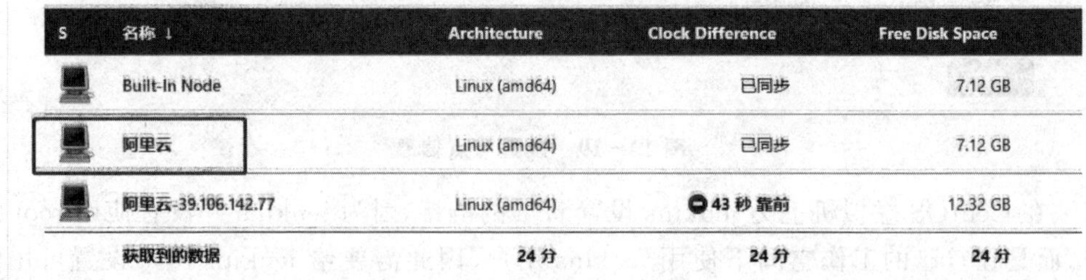

图 13-22 节点增加成功状态

13.6.2 为 Jenkins 配置 git 和 Allure 工具位置

一般,当 Jenkins 使用推荐的安装插件时,Git 插件会自动安装好,这里还需要再安装一个 Allure 插件,我们在 Jenkins 首页单击【系统管理】→【插件管理】按钮,如图 13-23 所示。

图 13 – 23　单击系统管理-插件管理

然后,切换到【可选插件】,搜索 Allure,选中 Allure 插件,再单击【Install Without restart】按钮,如图 13 – 24 所示。

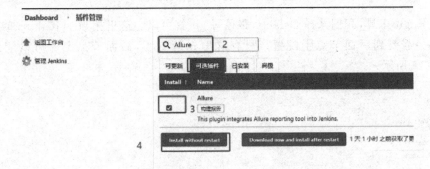

图 13 – 24　安装 allure 插件

选中【安装完成后重启 Jenkins】选项,安装完成,自动重启 Jenkins,之后重启 Jenkins,Allure 插件就安装成功了,如图 13 – 25 所示。

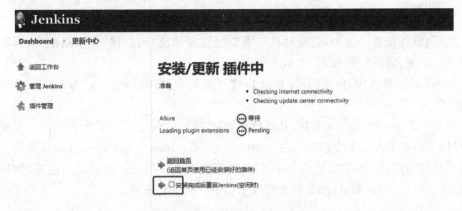

图 13 – 25　设置安装插件后重启

然后,在 Jenkins 首页,单击【系统管理】→【全局工具配置】,如图 13-26 所示。

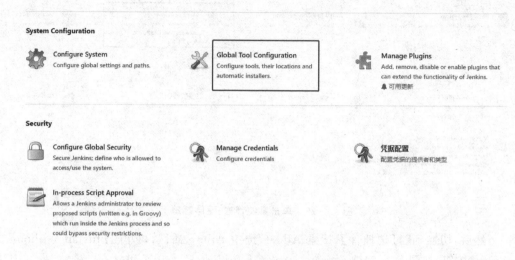

图 13-26　打开全局工具配置

配置 git 工具,我们选择 Default,路径写 git 就可以。这里相当于 Jenkins 的配置,具体路径需要到后面节点上配置。只有这里先配置了,后面节点上才可以配置,如图 13-27 所示。

图 13-27　配置全局 git 工具

之后,继续配置 Allure,这里同样不需要配置具体路径,同样,先全局配置,再到后面节点上配置具体路径,如图 13-28 所示。

在 Jenkins 首页中,单击【系统管理】→【节点管理】按钮,然后,单击对应的节点设置,如图 13-29 所示。

接着配置 git 和 Allure 工具,具体路径如图 13-30 所示。如果大家忘记了安装路径,可以通过在 CentOS 虚拟机上执行 which git 和 which allure 即可查询到路径。这里需要注意,Allure 不能配置到 bin 下,只能配置到 bin 的上一级目录。此外,如果不配置 JAVA_HOME 变量,那么就要配置 JAVA_HOME 变量。

Allure Commandline

Allure Commandline 安装

[新增 Allure Commandline]

Allure Commandline

别名

`allure`

安装目录

`allure`

⚠ allure is not a directory on the Jenkins controller (but perhaps it exists on some agents)

☐ Install automatically

图 13 – 28　配置 allure 工具

S	名称 ↓	Architecture	Clock Difference	Free Disk Space	Free Swap Space	Free Temp Space	Response Time	
💻	Built-In Node	Linux (amd64)	已同步	7.12 GB	⊖ 0 B	7.12 GB	0ms	⚙
💻	阿里云	Linux (amd64)	已同步	7.12 GB	⊖ 0 B	7.12 GB	32ms	⚙
💻	阿里云-39.106.142.77	Linux (amd64)	⊖ 43 秒 慢的	12.32 GB	⊖ 0 B	12.32 GB	31ms	⚙
	获取到的数据	28 分	28 分	28 分	28 分	28 分	28 分	

图 13 – 29　单击节点的设置

Dashboard ▸ Nodes ▸ 阿里云

Keep this agent online as much as possible

节点属性

☑ Environment variables

键值对列表

键

`JAVA_HOME`

值

`/usr/local/jdk1.8.0_331`

[新增]

☑ Tool Locations

工具位置列表

名称

`(Git) Default`

目录

`/usr/local/git/bin/git`

名称

`(Allure Commandline) allure`

目录

`/usr/local/allure-2.16.1`

图 13 – 30　配置 git 和 allure

13.7 基于Jenkins创建构建任务并生成Allure报告

首先,在Jenkins首页中,单击【新建Item】按钮,如图13-31所示。

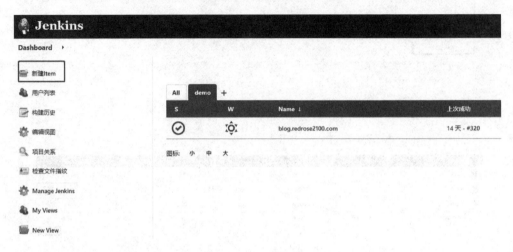

图 13-31 新建 Item

然后,设置任务名称,如 eTest,之后选择"Freestyle Project"类型,单击"OK",如图 13-32 所示。

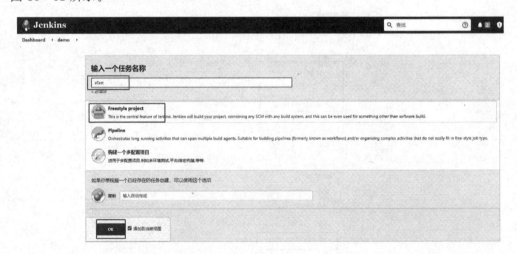

图 13-32 设置任务名称

从 github 仓库中复制链接,如图 13-33 所示。

将从 github 中复制的链接填入 Jenkins 中,然后单击"添加"—"Jenkins",配置秘钥,如图 13-34 所示。

图 13-33 从 github 中复制链接

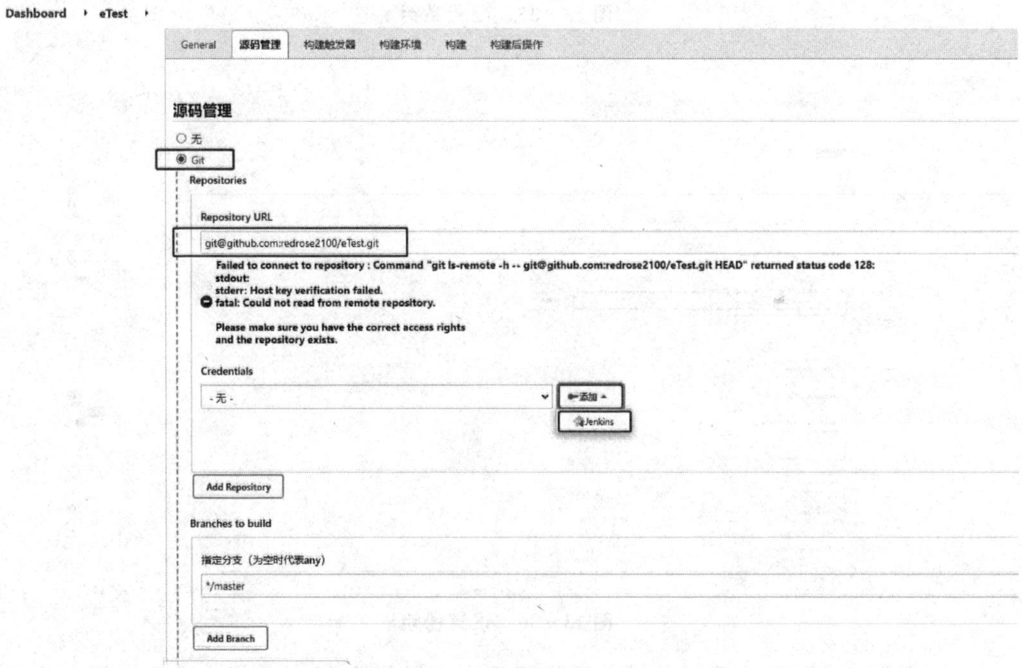

图 13-34 配置 git 代码仓 url

选择"SSH username with private key"类型,描述自定义内容,用于和其他密钥区分。然后填写 github 的用户名。这里 Private Key 为 CentOS 节点上 id_rsa 文件的私钥内容。如图 13-35 所示。

如图 13-36,回到配置任务的页面,选择刚刚配置的密钥,可以看到,此时报错信息已经消失了,选择 master 分支。

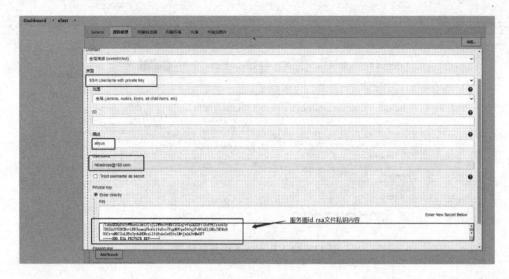

图 13-35 配置私钥

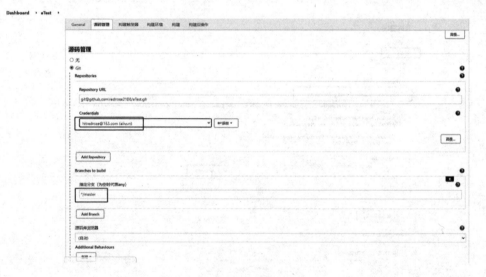

图 13-36 设置秘钥

此外,在配置 Git 代码的上面还需要配置执行任务的节点,我们这里的标签表达式为节点配置上设置的标签,是 aliyun,如图 13-37 所示。

然后,单击"增加构建步骤",选择"Execute Shell",如图 13-38 所示。

在 shell 中编写如下命令行。

```
pip install -r requirements.txt
cd eTest/tests/
python3 -m pytest --alluredir reports/allure_report
```

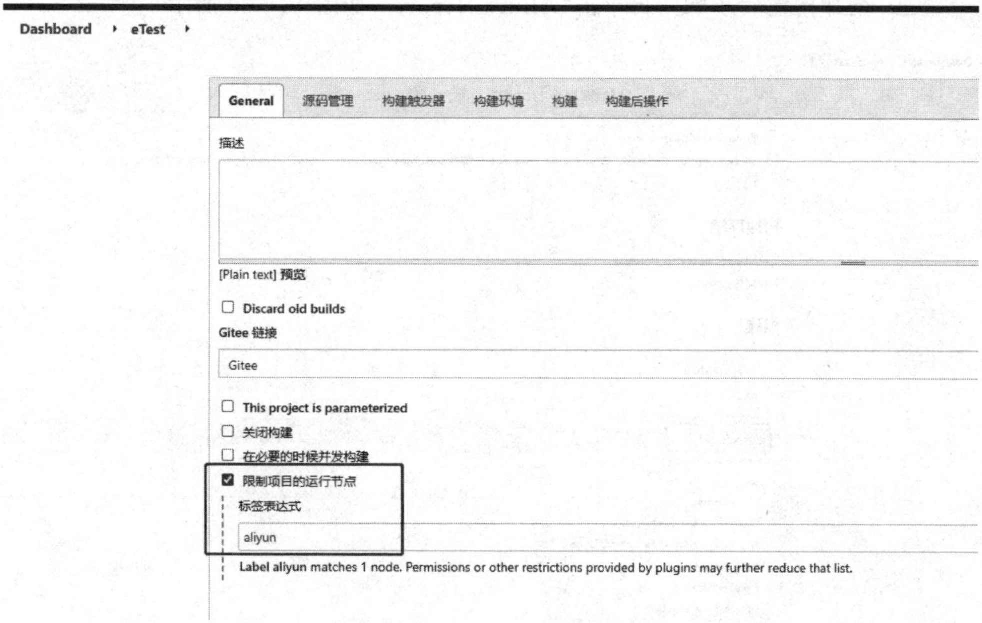

图 13-37　设置执行任务的节点

图 13-38　增加构建步骤

单击"增加构建后步骤",再单击"Allure Report",如图13-39所示。

图 13-39 增加构建后步骤

设置 Allure Report 中的 Path 与 shell 中命令行生成的路径为一致,如这里都设置为 reports/allure_report,需要注意,Path 要从项目的根目录开始计算,因为 shell 中有切换目录的操作,因此 Path 中需要将 shell 中切换的目录也一起计算上,如图13-40所示。

保存后,单击"构建",即开始执行,如图13-41所示。

如图13-42所示,执行成功后,单击 Allure Report,即可查看 Allure 报告。

Allure 报告如图13-43所示,可以看出,Allure 自动化测试报告。脚本中的功能、用户故事,用例名称、测试步骤等都显示出来了。日志打印部分在 log 部分显示出来。

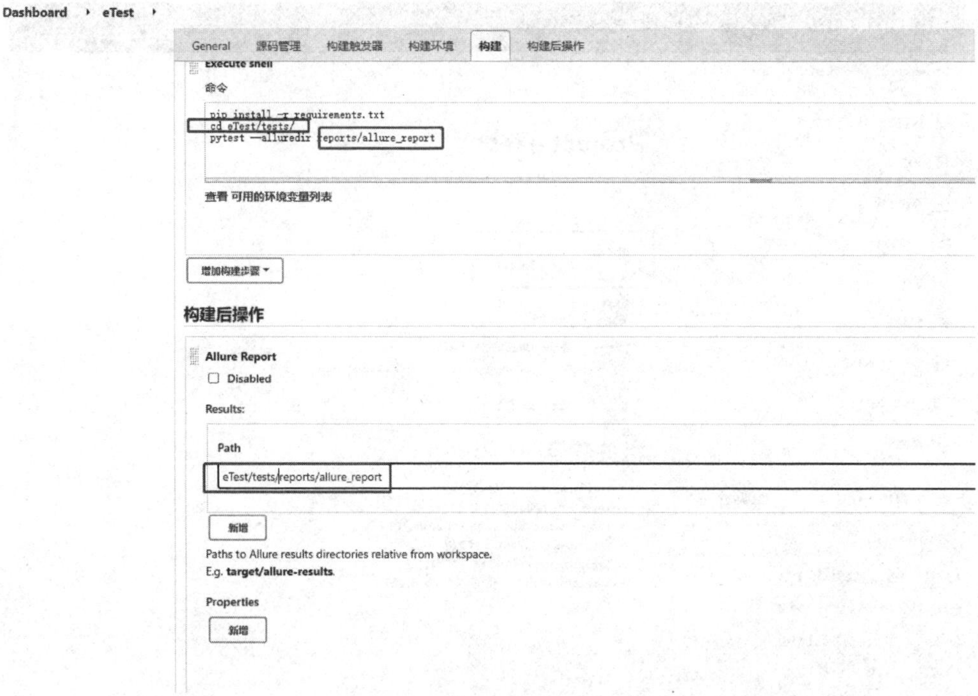

图 13-40 设置 Allure Report

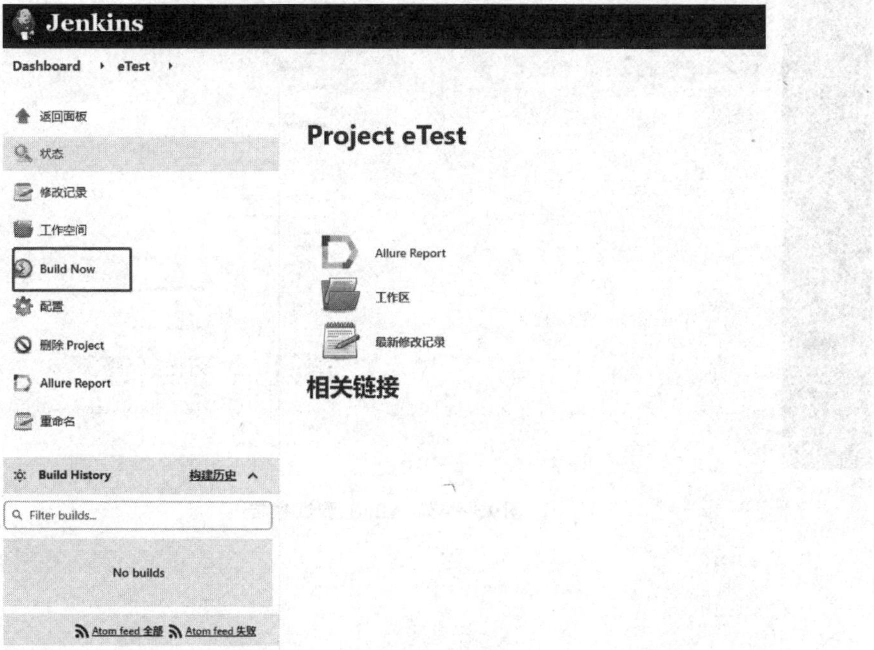

图 13-41 开始执行构建任务

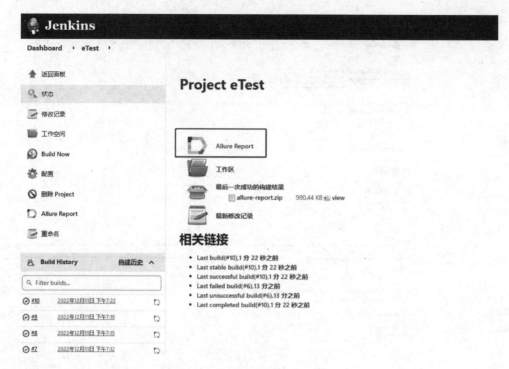

图 13-42　单击查看 allure 报告

图 13-43　Allure 测试报告

第 14 章 其他常用的用例插件

14.1 Pytest-attrib 根据属性挑选用例

Pytest-attrib 件是提供根据属性挑选用例的,类似于 mark 的功能。Pytest-attrib 用在测试类中挑选用例是非常方便的,其不使用装饰器,即可根据测试类的属性进行挑选,下面我们首先使用如下命令安装 Pytest-attrib 插件。

```
pip install pytest-attrib
```

为了更好显示 Pytest-attrib 插件的用处,这里准备了三个测试类,其中,TestDemo 类中有 smoke 属性,TestDemo2 中有 feature 属性,而 TestDemo3 类中既有 smoke 又有 feature 属性。

```python
class TestDemo(object):
    smoke = True
    def test_01(self):
        print("in TestDemo.test_01...")

    def test_02(self):
        print("in TestDemo.test_02...")


class TestDemo2(object):
    feature = True

    def test_03(self):
        print("in TestDemo2.test_03...")

    def test_04(self):
        print("in TestDemo2.test_04...")

class TestDemo3(object):
    feature = True
    smoke = True
```

```python
    def test_05(self):
        print("in TestDemo3.test_05 ...")

    def test_06(self):
        print("in TestDemo3.test_06 ...")
```

如想执行带有 smoke 属性的用例，只需使用-a 参数指定 smoke 即可，如下所示。

可以看出，此时因为 TestDemo 类和 TestDemo3 类中均有 smoke 属性，因此，两个类中的测试方法均执行。

```
(demo-HCIhX0Hq) E:\demo> pytest -s -a smoke
==================== test session starts ====================
platform win32 -- Python 3.7.9, pytest-7.2.0, pluggy-1.0.0
rootdir: E:\demo, configfile: pytest.ini
plugins: assume-2.4.3, attrib-0.1.3, rerunfailures-10.2
collected 6 items / 2 deselected / 4 selected

test_demo.py in TestDemo.test_01 ...
.in TestDemo.test_02 ...
.in TestDemo3.test_05 ...
.in TestDemo3.test_06 ...
.

============= 4 passed, 2 deselected in 0.03s =============

(demo-HCIhX0Hq) E:\demo>
```

如果想执行既有 smoke 属性又有 feature 属性的测试类用例，只需要使用 and 逻辑词连接即可。这里同时拥有 smoke 和 feature 属性的类 TestDemo3，因此将 TestDemo3 类中的所有测试方法均执行，如下所示。

```
(demo-HCIhX0Hq) E:\demo> pytest -s -a "smoke and feature"
==================== test session starts ====================
platform win32 -- Python 3.7.9, pytest-7.2.0, pluggy-1.0.0
rootdir: E:\demo, configfile: pytest.ini
plugins: assume-2.4.3, attrib-0.1.3, rerunfailures-10.2
collected 6 items / 4 deselected / 2 selected

test_demo.py in TestDemo3.test_05 ...
.in TestDemo3.test_06 ...
.

============= 2 passed, 4 deselected in 0.03s =============
```

(demo-HCIhX0Hq) E:\demo>

如果想执行带有 smoke 属性的或者带有 feature 属性的测试类脚本,只需要使用 or 连接词连接即可。这里带有 smoke 属性或者带有 feature 属性的测试类有 TestDemo、TestDemo2、TestDemo3,因此,将此三个类中的所有测试方法均执行了,如下所示。

```
(demo-HCIhX0Hq) E:\demo> pytest -s -a "smoke or feature"
=================== test session starts ===================
platform win32 -- Python 3.7.9, pytest-7.2.0, pluggy-1.0.0
rootdir: E:\demo, configfile: pytest.ini
plugins: assume-2.4.3, attrib-0.1.3, rerunfailures-10.2
collected 6 items

test_demo.py in TestDemo.test_01 ...
. in TestDemo.test_02 ...
. in TestDemo2.test_03 ...
. in TestDemo2.test_04 ...
. in TestDemo3.test_05 ...
. in TestDemo3.test_06 ...
.

=================== 6 passed in 0.03s ===================

(demo-HCIhX0Hq) E:\demo>
```

如果想执行不带 smoke 属性的用例,则需要使用逻辑关系词 not 来处理。从下面代码中可以看出,不带 smoke 属性的测试类只有 TestDemo2,因此只执行了 TestDemo2 中的测试方法。

```
(demo-HCIhX0Hq) E:\demo> pytest -s -a "not smoke"
=================== test session starts ===================
platform win32 -- Python 3.7.9, pytest-7.2.0, pluggy-1.0.0
rootdir: E:\demo, configfile: pytest.ini
plugins: assume-2.4.3, attrib-0.1.3, rerunfailures-10.2
collected 6 items / 4 deselected / 2 selected

test_demo.py in TestDemo2.test_03 ...
. in TestDemo2.test_04 ...
.

============= 2 passed, 4 deselected in 0.02s =============

(demo-HCIhX0Hq) E:\demo>
```

14.2　Pytest-sugar 执行过程中显示进度条

Pytest-sugar 是一款用来改善控制台显示的插件,其增加了进度条显示,使得在用例执行过程中可以看到进度条。进度条根据用例是否通过标注了不同颜色,如用例通过标记为绿色,用例失败则标记为红色,非常醒目。

我们先使用如下命令安装 Pytest-sugar 插件。

```
pip install Pytest-sugar
```

然后,准备如下测试脚本。

```python
def test_1():
    print("in test_1")
    assert 1 == 1
def test_2():
    print("in test_2")
    assert 1 == 1
def test_3():
    print("in test_3")
    assert 1 == 1
def test_4():
    print("in test_4")
    assert 1 == 1
def test_5():
    print("in test_5")
    assert 1 == 1
def test_6():
    print("in test_6")
    assert 1 == 1
```

执行 Pytest 命令结果如下所示,可以发现,在 test_demo.py 后面跟着打印了若干个对号,同时有一段条形图,控制台根据用例是否通过显示了不同的颜色。

```
(demo-HCIhX0Hq) E:\demo> pytest
Test session starts (platform: win32, Python 3.7.9, pytest 7.2.0, pytest-sugar 0.9.6)
benchmark: 4.0.0 (defaults: timer=time.perf_counter disable_gc=False min_rounds=5 min_time=0.000005 max_time=1.0 calibration_precision=10 warmup=False warmup_iterations=100000)
rootdir: E:\demo, configfile: pytest.ini
plugins: allure-pytest-2.12.0, assume-2.4.3, attrib-0.1.3, benchmark-4.0.0, rerunfailures-10.2, sugar-0.9.6
collecting ...
```

```
test_demo.py ✓✓✓✓✓                                    100%
Results (0.10s):
       6 passed

(demo - HCIhX0Hq) E:\demo>
```

如果安装了 Pytest-sugar 后,在默认情况下就是开启的,若不想使用 Pytest-sugar 了,可以直接卸载或者使用命令禁止使用 Pytest-sugar,如下所示。

```
(demo - HCIhX0Hq) E:\demo> pytest -p no:sugar
===================== test session starts =====================
platform win32 -- Python 3.7.9, pytest-7.2.0, pluggy-1.0.0
benchmark: 4.0.0 (defaults: timer=time.perf_counter disable_gc=False min_rounds=5 min_time=0.000005 max_time=1.0 calibration_precision=10 warmup=False warmup_iterations=100000)
rootdir: E:\demo, configfile: pytest.ini
plugins: allure-pytest-2.12.0, assume-2.4.3, attrib-0.1.3, benchmark-4.0.0, rerunfailures-10.2
collected 6 items

test_demo.py ......                                    [100%]

===================== 6 passed in 0.09s =====================

(demo - HCIhX0Hq) E:\demo>
```

14.3 Pytest-csv 执行结果输出 csv 文件

Pytest-csv 插件用于将自动化脚本执行结果存储为 csv 格式,我们知道 csv 格式数据是可以直接使用 excel 打开的,这样就可以将脚本执行结果直接存放于 excel 表格中,这是非常方便的。

我们先执行如下命令,安装 Pytest-csv 插件。

```
pip install pytest-csv
```

为了更好地演示 Pytest-csv 插件的使用方法以及效果,这里准备了如下测试脚本。

```
import pytest


@pytest.mark.unit
@pytest.mark.smoke
def test_1():
    """
```

```
测试用例 1：用于演示 Pytest-csv 插件的作用
:return:
"""
print("in test_1")
assert 1 == 1


@pytest.mark.system
def test_2():
    """
    测试用例 2：用于演示 Pytest-csv 插件的作用
    :return:
    """
    print("in test_2")
    assert 1 == 2
```

然后，使用命令执行脚本，此时即会自动将脚本执行结果以及脚本存储为 csv 格式。

```
pytest --csv test.csv
```

再使用 excel 表格打开 csv 格式的数据，结果如图 14-1 所示，这里在表格中将用例 id、模块名、函数名、文件名、注释、标签、执行结果、报错信息以及延时等均显示出来了，相当于自动生成了 excel 数据表格式的自动化测试报表。这对测试来说是非常方便的。

id	module	name	file	doc	markers	status	message	duration
test_demo.py::test_1	test_demo	test_1	test_demo.py	测试用例1: 用于演示pytest-csv 插件的作用 :return:	smoke,unit	passed		0.000436
test_demo.py::test_2	test_demo	test_2	test_demo.py	测试用例2: 用于演示pytest-csv 插件的作用 :return:	system	failed	assert 1 == 2	0.000598

图 14-1 Pytest-csv 生成的数据格式

14.4 用 Pytest-tldr 插件简化脚本执行日志输出

Pytest 测试框架在执行的时候多实时标准输出、实时日志、捕获日志、捕获标准输出等多种类型的内容在控制台显示，许多人觉得这种输出不是太友好，尤其是当用例失败后，来回上下翻控制台的输出却不易找到报错位置。而 Pytest-tldr 插件就是为了解决这个问题的，其用例报错后仅显示报错用例的调用错误栈，其他不显示。

为了更好体验 Pytest-tldr 插件的效果，在安装 Pytest-tldr 之前，我们先利用如下用例执行一次，观察一下报错信息。

```
def test_1():
    print("in test_1")
    assert 1 == 1

def test_2():
    print("in test_2")
    assert 1 == 2
```

从执行结果中可以看到，其实这个用例非常简单，是直接判断 1 和 2 是否相等的断言，但是经 Pytest 执行后打印出来的信息却很多，使大家，尤其是新手会感觉无法入手定位问题。

```
(demo-HCIhXOHq) E:\demo> pytest
Test session starts (platform: win32, Python 3.7.9, pytest 7.2.0, pytest-sugar 0.9.6)
benchmark: 4.0.0 (defaults: timer=time.perf_counter disable_gc=False min_rounds=5 min_time=0.000005 max_time=1.0 calibration_precision=10 warmup=False warmup_iterations=100000)
rootdir: E:\demo, configfile: pytest.ini
plugins: allure-pytest-2.12.0, assume-2.4.3, attrib-0.1.3, benchmark-4.0.0, csv-3.0.0, rerunfailures-10.2, sugar-0.9.6
collecting ...
 test_demo.py ✓                                      50%

―――――――――――――――――――――― test_2 ――――――――――――――――――――――

    def test_2():
        print("in test_2")
>       assert 1 == 2
E       assert 1 == 2

test_demo.py:8: AssertionError
------------------- Captured stdout call -------------------
in test_2

 test_demo.py ×                                      100%
============= short test summary info =============
FAILED test_demo.py::test_2 - assert 1 == 2

Results (0.15s):
       1 passed
       1 failed
```

341

```
- test_demo.py:6 test_2

(demo - HCIhXOHq) E:\demo>
```

之后,通过如下命令安装 Pytest-tldr 插件。

```
pip install pytest-tldr
```

安装完成后,再次使用 Pytest 执行,可以看出,此时的结构就简单多了,很明了。

```
(demo - HCIhXOHq) E:\demo> pytest
.F
==================================================================
FAIL: test_demo.py::test_2
------------------------------------------------------------------
in test_2

Traceback (most recent call last):
  File "E:\demo\test_demo.py", line 8, in test_2
    assert 1 == 2
AssertionError: assert 1 == 2

------------------------------------------------------------------
Ran 2 tests in 0.08s

FAILED (failures = 1)

(demo - HCIhXOHq) E:\demo>
```

将用例断言改为成功,如下所示。

```
def test_1():
    print("in test_1")
    assert 1 == 1

def test_2():
    print("in test_2")
    assert 1 == 1
```

关闭 tldr 插件,执行结果如下所示。

```
(demo - HCIhXOHq) E:\demo> pytest - p no:tldr
Test session starts (platform: win32, Python 3.7.9, pytest 7.2.0, pytest - sugar 0.9.6)
benchmark: 4.0.0 (defaults: timer = time.perf_counter disable_gc = False min_rounds = 5 min_time = 0.000005 max_time = 1.0 calibration_precision = 10 warmup = False warmup_iterations = 100000)
```

rootdir: E:\demo, configfile: pytest.ini
plugins: allure-pytest-2.12.0, assume-2.4.3, attrib-0.1.3, benchmark-4.0.0, csv-3.0.0, rerunfailures-10.2, sugar-0.9.6
collecting ...
 test_demo.py ✓✓   100%

Results (0.09s):
       2 passed

(demo-HCIhX0Hq) E:\demo>

放开tldr插件，可以发现，此时显示非常简洁。

(demo-HCIhX0Hq) E:\demo> pytest
..

Ran 2 tests in 0.08s

OK

(demo-HCIhX0Hq) E:\demo>

第 15 章 Pytest 核心即 Pluggy 源码解读

15.1 Pluggy 模块的应用

简单讲，Pluggy 模块就是一个插件系统，Pytest 自动化测试框架的核心就是 Pluggy，Pytest 就是基于 Pluggy 的将一系列插件组装在一起的测试框架，因此，这里首先介绍 pluggy 模块的使用方法。

首先安装 Pluggy 模块，执行如下命令。

```
pip install pluggy
```

然后，编写实例代码，这里面主要有三个类，其中 MySpec 类可以简单理解为接口类，其中，Plugin_1 和 Plugin_2 可以理解为实现类，也可以理解为插件，其对同一个接口可以有多个实现，在主函数中，pm 可以理解为一个插件管理器，即 pm.add_hookspecs 用于为插件系统增加接口，pm.register 则是插件注册的过程。当插件注册后，pm.hook.myhook 会执行 myhook 函数。当有多个插件注册而且有多个插件中都实现了接口函数后，每个插件的接口函数都会执行，且按照后注册先执行的顺序将所有注册了的插件方法统一都执行一遍。

```
import pluggy

hookspec = pluggy.HookspecMarker("myproject")
hookimpl = pluggy.HookimplMarker("myproject")

class MySpec:
    @hookspec
    def myhook(self, arg1, arg2):
        pass

class Plugin_1:
    @hookimpl
    def myhook(self, arg1, arg2):
        print("inside Plugin_1.myhook()")
        return arg1 + arg2
```

```
class Plugin_2:
    @hookimpl
    def myhook(self, arg1, arg2):
        print("inside Plugin_2.myhook()")
        return arg1 - arg2

if __name__ == "__main__":
    pm = pluggy.PluginManager("myproject")
    pm.add_hookspecs(MySpec)
    pm.register(Plugin_1())
    pm.register(Plugin_2())
    results = pm.hook.myhook(arg1=1, arg2=2)
    print(results)
```

执行结果如下所示,可以看到,这里先执行了 Plugin_2 中的 myhook 方法,然后又执行了 Plugin_1 中的 myhook 方法,之后按照此顺序将每个插件中的方法返回值放在列表中做了返回。

```
(demo-HCIhXOHq) E:\demo> python demo.py
inside Plugin_2.myhook()
inside Plugin_1.myhook()
[-1, 3]

(demo-HCIhXOHq) E:\demo>
```

下面再看一个更为复杂一点的实例,如下所示。这里可以理解为 MySpec 类定义了四个接口函数,分别是 myhook1、myhook2、myhook3、myhook4,然后又定义了两个插件类,插件 Pluggin_1 实现了 myhook1、myhook2、myhook3,而插件 Pluggin_2 实现 myhook2、myhook3、myhook4。

```
import pluggy

hookspec = pluggy.HookspecMarker("myproject")
hookimpl = pluggy.HookimplMarker("myproject")

class MySpec:
    @hookspec
    def myhook1(self, arg1, arg2):
        pass
    @hookspec
    def myhook2(self, arg1, arg2):
        pass
    @hookspec
    def myhook3(self, arg1, arg2):
        pass
```

```python
    @hookspec
    def myhook4(self, arg1, arg2):
        pass

class Plugin_1:
    @hookimpl
    def myhook1(self, arg1, arg2):
        print("inside Plugin_1.myhook1()")
        return arg1 + arg2
    @hookimpl
    def myhook2(self, arg1, arg2):
        print("inside Plugin_1.myhook2()")
        return arg1 + arg2 + 1
    @hookimpl
    def myhook3(self, arg1, arg2):
        print("inside Plugin_1.myhook4()")
        return arg1 + arg2 + 2


class Plugin_2:

    @hookimpl
    def myhook2(self, arg1, arg2):
        print("inside Plugin_2.myhook2()")
        return arg1 - arg2
    @hookimpl
    def myhook3(self, arg1, arg2):
        print("inside Plugin_2.myhook3()")
        return arg1 - arg2 - 1


    @hookimpl
    def myhook4(self, arg1, arg2):
        print("inside Plugin_2.myhook4()")
        return arg1 - arg2 - 2

if __name__ == "__main__":
    pm = pluggy.PluginManager("myproject")
    pm.add_hookspecs(MySpec)
    pm.register(Plugin_1())
    pm.register(Plugin_2())
    results = pm.hook.myhook1(arg1=1, arg2=2)
    print(results)
    results = pm.hook.myhook2(arg1=1, arg2=2)
```

```
    print(results)
    results = pm.hook.myhook3(arg1 = 1, arg2 = 2)
    print(results)
    results = pm.hook.myhook4(arg1 = 1, arg2 = 2)
    print(results)
```

执行结果如下所示,因为 myhook1 只有一个插件实现,因此,返回值只有一个,myhook2 和 myhook3 分别有两个插件实现,返回值有两个,且按照先进后出的顺序存入列表,同样,myhook4 也只有一个插件定义,因此也只有一个返回值。

```
(demo-HCIhX0Hq) E:\demo> python demo.py
inside Plugin_1.myhook1()
[3]
inside Plugin_2.myhook2()
inside Plugin_1.myhook2()
[-1, 4]
inside Plugin_2.myhook3()
inside Plugin_1.myhook4()
[-2, 5]
inside Plugin_2.myhook4()
[-3]

(demo-HCIhX0Hq) E:\demo>
```

在定义接口时,可以通过 firstresult = True 指定:只要有一个结果正确返回了,就结束。在下面代码中,定义 myhook 时通过装饰器指定了 firstresult = True,这样在执行的时候,会注册 Pluggin_1 和 Pluggin_2 两个插件。仍然按照后进先出的顺序,先执行 Pluggin_2 插件,返回正确,那么就不会再执行 Plugging_1 中的 myhook 了。当然,这是通过代码分析出来的,下面通过执行结果验证一下。

```
import pluggy

hookspec = pluggy.HookspecMarker("myproject")
hookimpl = pluggy.HookimplMarker("myproject")

class MySpec:
    @hookspec(firstresult = True)
    def myhook(self, arg1, arg2):
        pass

class Plugin_1:
    @hookimpl
    def myhook(self, arg1, arg2):
        print("in Plugin_1.myhook()")
```

```python
        return arg1 + arg2

class Plugin_2:
    @hookimpl
    def myhook(self, arg1, arg2):
        print("in Plugin_2.myhook()")
        return arg1 - arg2

pm = pluggy.PluginManager("myproject")
pm.add_hookspecs(MySpec)
pm.register(Plugin_1())
pm.register(Plugin_2())
results = pm.hook.myhook(arg1 = 1, arg2 = 2)
print(results)
```

从执行结果中可以看到,这里只执行了 Plugin_2 类中的 myhook 方法,Plugin_1 类中的 myhook 不再执行。

```
(demo-HCIhXOHq) E:\demo> python demo.py
in Plugin_2.myhook()
-1
```

此外,我们还可以通过 tryfirst 或 trylast 指定当前类最先执行或者最后执行如下代码所示。

```python
import pluggy

hookspec = pluggy.HookspecMarker("myproject")
hookimpl = pluggy.HookimplMarker("myproject")

class MySpec:
    @hookspec
    def myhook(self, arg1, arg2):
        pass

class Plugin_1:
    @hookimpl
    def myhook(self, arg1, arg2):
        print("in Plugin_1.myhook()")
        return arg1 + arg2

class Plugin_2:
    @hookimpl(tryfirst = True)
    def myhook(self, arg1, arg2):
```

```
        print("in Plugin_2.myhook()")
        return arg1 - arg2

class Plugin_3:
    @hookimpl
    def myhook(self, arg1, arg2):
        print("in Plugin_3.myhook()")
        return arg1 - arg2 + 10

pm = pluggy.PluginManager("myproject")
pm.add_hookspecs(MySpec)
pm.register(Plugin_1())
pm.register(Plugin_2())
pm.register(Plugin_3())
results = pm.hook.myhook(arg1 = 1, arg2 = 2)
print(results)
```

可以看出,此时首先执行了 Plugin_2 类。

```
(demo-HCIhXOHq) E:\demo> python demo.py
in Plugin_2.myhook()
in Plugin_3.myhook()
in Plugin_1.myhook()
[-1, 9, 3]

(demo-HCIhXOHq) E:\demo>
```

同样,我们可以通过 trylast 指定当前类最后执行,比如这里把 Plugin_2 中的 tryfirst 修改为 trylast,再次执行的结果如下所示:

```
(demo-HCIhXOHq) E:\demo> python demo.py
in Plugin_3.myhook()
in Plugin_1.myhook()
in Plugin_2.myhook()
[9, 3, -1]

(demo-HCIhXOHq) E:\demo>
```

我们看,当传入 hookwrapper=True 时,需要在 Plugin 中实现一个 yield,Plugin 先执行 yield 之前的代码,再执行其他的 Pluggin,然后回来执行 yield 之后的代码。同时通过 yield 可以获取其他插件执行的结果。如下所示,在 Plugin_2 中使用了 yield,在 yield 执行完成后可以获取所有插件的执行结果为 output。同时,当所有 yield 用的代码执行完成之后,仍然可以获取所有插件的返回值列表。

```
import pluggy
```

```python
hookspec = pluggy.HookspecMarker("myproject")
hookimpl = pluggy.HookimplMarker("myproject")

class MySpec:
    @hookspec
    def myhook(self, arg1, arg2):
        pass

class Plugin_1:
    @hookimpl()
    def myhook(self, arg1, arg2):
        print("in Plugin_1.myhook()")
        return arg1 + arg2

class Plugin_2:
    @hookimpl(hookwrapper=True)
    def myhook(self, arg1, arg2):
        print("in Plugin_2.myhook() before yield...")
        output = yield
        result = output.get_result()
        print("in Plugin_2.myhook() after yield...")
        print(result)


class Plugin_3:
    @hookimpl
    def myhook(self, arg1, arg2):
        print("inside Plugin_3.myhook()")
        return arg1 - arg2 + 10

pm = pluggy.PluginManager("myproject")
pm.add_hookspecs(MySpec)
pm.register(Plugin_1())
pm.register(Plugin_2())
pm.register(Plugin_3())
results = pm.hook.myhook(arg1=1, arg2=2)
print("after all run...")
print(results)
```

执行结果如下,可以看到在 Plugin_2 中 yield 语句之后以及在所有代码执行完成后,都可以获取所有插件的返回值。

```
(demo-HCIhXOHq) E:\demo> python demo.py
```

```
in Plugin_2.myhook() before yield...
inside Plugin_3.myhook()
in Plugin_1.myhook()
in Plugin_2.myhook() after yield...
[9, 3]
after all run ...
[9, 3]

(demo - HCIhXOHq) E:\demo >
```

15.2 Pluggy 源码解读基础准备

解读 Pluggy 源码，直接使用 Pytest 环境中安装的 Pluggy 即可完成，我们这里安装的 Pluggy 版本是 1.0，为了让大家更好理解源码，我们首先使用代码作为实例为大家介绍。Pluggy 的使用步骤主要包括以下几个步骤：

（1）将 HookspecMarker 类实例化。
（2）将 HookimplMarker 类实例化。
（3）定义一个接口类，将接口类中的函数上使用(1)中的实例化作为装饰器装饰。
（4）定义一个实现类，将实现类中的函数上使用(2)中的实例化作为装饰器装饰。
（5）将 PluginManager 类实例化。
（6）对(5)中的实例调用 add_hookspecs 方法。
（7）使用(5)中的实例调用注册方法。
（8）通过(5)中实例的 hook 属性调用接口方法。

```python
import pluggy

hookspec = pluggy.HookspecMarker("myproject")
hookimpl = pluggy.HookimplMarker("myproject")

class MySpec:
    @hookspec
    def myhook(self, arg1, arg2):
        pass

class Plugin_1:
    @hookimpl()
    def myhook(self, arg1, arg2):
        print("in Plugin_1.myhook()")
        return arg1 + arg2
```

```python
class Plugin_2:
    @hookimpl(hookwrapper = True)
    def myhook(self, arg1, arg2):
        print("in Plugin_2.myhook() before yield...")
        output = yield
        result = output.get_result()
        print("in Plugin_2.myhook() after yield...")
        print(result)


class Plugin_3:
    @hookimpl
    def myhook(self, arg1, arg2):
        print("inside Plugin_3.myhook()")
        return arg1 - arg2 + 10


pm = pluggy.PluginManager("myproject")
pm.add_hookspecs(MySpec)
pm.register(Plugin_1())
pm.register(Plugin_2())
pm.register(Plugin_3())
results = pm.hook.myhook(arg1 = 1, arg2 = 2)
print("after all run ...")
print(results)
```

为更好地定位跳转，我们可以在 pycharm 中打开上述测试代码，并配置解释器，然后即可在 External Library 中找到 Pluggy 的源码了，如图 15-1 所示。

我们发现 pluggy 源码只有以下 7 个文件。

```
pluggy
    |-------- __init__.py
    |-------- _caller.py
    |-------- _hooks.py
    |-------- _manager.py
    |-------- _result.py
    |-------- _tracing.py
    |-------- _version.py
```

其中 __init__.py 文件代码如下所示，即这里通过 __all__ 限定了 Pluggy，外部只允许使用 PluginManager、PluginValidationError、HookCallError、HookspecMarker、HookimplMarker 这五个类。

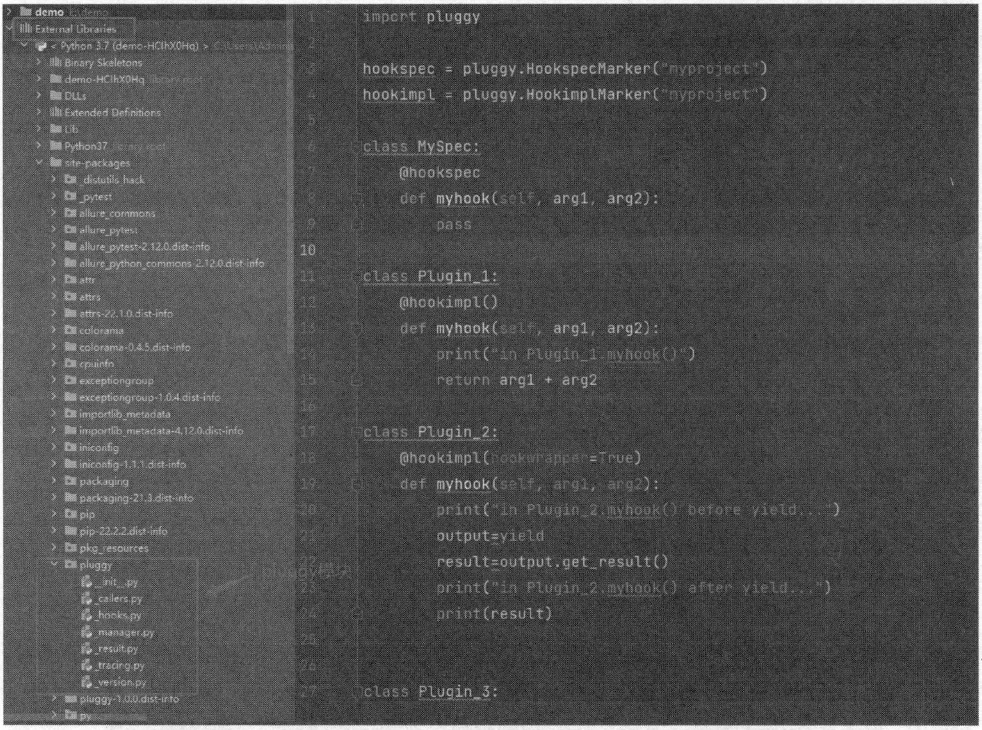

图 15-1 pycharm 中显示 pluggy 源码

```python
try:
    from ._version import version as __version__
except ImportError:
    # broken installation, we don't even try
    # unknown only works because we do poor mans version compare
    __version__ = "unknown"


__all__ = [
    "PluginManager",
    "PluginValidationError",
    "HookCallError",
    "HookspecMarker",
    "HookimplMarker",
]

from ._manager import PluginManager, PluginValidationError
from ._callers import HookCallError
from ._hooks import HookspecMarker, HookimplMarker
```

_version.py 中的内容如下所示,这里设置的是 Pluggy 版本号。

```
# coding: utf-8
# file generated by setuptools_scm
# don't change, don't track in version control
version = '1.0.0'
version_tuple = (1, 0, 0)
```

然后,在 pycharm 中对 Pluggy 应用实例脚本中使用 Ctrl 键,跳转到对应的函数定义,如图 15-2 所示。

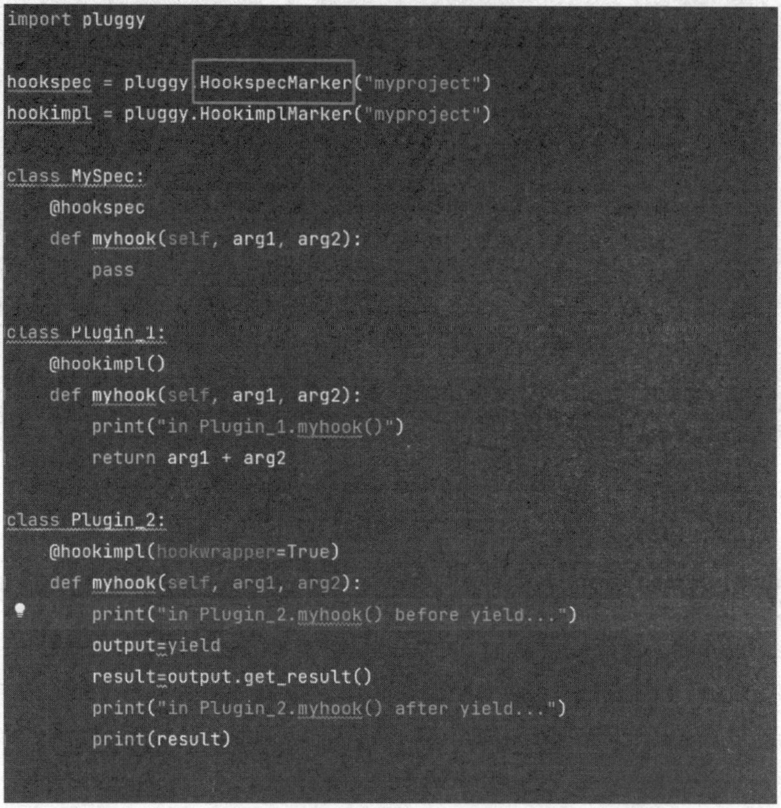

```
import pluggy

hookspec = pluggy.HookspecMarker("myproject")
hookimpl = pluggy.HookimplMarker("myproject")

class MySpec:
    @hookspec
    def myhook(self, arg1, arg2):
        pass

class Plugin_1:
    @hookimpl()
    def myhook(self, arg1, arg2):
        print("in Plugin_1.myhook()")
        return arg1 + arg2

class Plugin_2:
    @hookimpl(hookwrapper=True)
    def myhook(self, arg1, arg2):
        print("in Plugin_2.myhook() before yield...")
        output=yield
        result=output.get_result()
        print("in Plugin_2.myhook() after yield...")
        print(result)
```

图 15-2 按住 Ctrl 键,单击鼠标即可跳转到定义的位置

然后,就可以根据 Pluggy 的应用实例代码对 Pluggy 源码进行解析了。

15.3 HookspecMarker 类和 HookimplMarker 类分析

从 Pluggy 模块应用方法实例可以看出,HookspecMarker 类和 HookimplMarker 类首先对进行了实例化,下面我们解读一下这两个类的源码。

HookspecMarker 类和 HookimplMarker 的定义如下所示,我们先看 Hookspec-Marker 类的定义,类中只有一个 __init__ 函数和 __call__ 函数,__init__ 函数就是一个赋

值的操作,即在 HookspecMarker 实例化的时候传入了一个名称,然后就有了一个 project_name 的属性,属性值即为传入的值。HookimplMarker 同样也只有一个 __init__ 函数和一个 __call__ 函数,__init__ 函数同样也是给实例 project_name 赋值的。

```
class HookspecMarker:
    """Decorator helper class for marking functions as hook specifications.

    You can instantiate it with a project_name to get a decorator.
    Calling :py:meth:`.PluginManager.add_hookspecs` later will discover all marked functions
    if the :py:class:`.PluginManager` uses the same project_name.
    """

    def __init__(self, project_name):
        self.project_name = project_name

    def __call__(
        self, function=None, firstresult=False, historic=False, warn_on_impl=None
    ):
        """if passed a function, directly sets attributes on the function
        which will make it discoverable to :py:meth:`.PluginManager.add_hookspecs`.
        If passed no function, returns a decorator which can be applied to a function
        later using the attributes supplied.

        If "firstresult" is "True" the 1:N hook call (N being the number of registered
        hook implementation functions) will stop at I<=N when the I'th function
        returns a non-"None" result.

        If "historic" is "True" calls to a hook will be memorized and replayed
        on later registered plugins.

        """

        def setattr_hookspec_opts(func):
            if historic and firstresult:
                raise ValueError("cannot have a historic firstresult hook")
            setattr(
                func,
                self.project_name + "_spec",
                dict(
                    firstresult=firstresult,
                    historic=historic,
                    warn_on_impl=warn_on_impl,
                ),
```

```python
            )
            return func

        if function is not None:
            return setattr_hookspec_opts(function)
        else:
            return setattr_hookspec_opts


class HookimplMarker:
    """ Decorator helper class for marking functions as hook implementations.

    You can instantiate with a "project_name" to get a decorator.
    Calling :py:meth:`.PluginManager.register` later will discover all marked functions
    if the :py:class:`.PluginManager` uses the same project_name.
    """

    def __init__(self, project_name):
        self.project_name = project_name

    def __call__(
        self,
        function=None,
        hookwrapper=False,
        optionalhook=False,
        tryfirst=False,
        trylast=False,
        specname=None,
    ):

        """ if passed a function, directly sets attributes on the function
        which will make it discoverable to :py:meth:`.PluginManager.register`.
        If passed no function, returns a decorator which can be applied to a
        function later using the attributes supplied.

        If ``optionalhook`` is ``True`` a missing matching hook specification will not result
        in an error (by default it is an error if no matching spec is found).

        If ``tryfirst`` is ``True`` this hook implementation will run as early as possible
        in the chain of N hook implementations for a specification.

        If ``trylast`` is ``True`` this hook implementation will run as late as possible
        in the chain of N hook implementations.
```

```
If "hookwrapper" is "True" the hook implementations needs to execute exactly
one "yield".  The code before the "yield" is run early before any non-hookwrapper
function is run.  The code after the "yield" is run after all non-hookwrapper
function have run.  The "yield" receives a :py:class:'.callers._Result' object
representing the exception or result outcome of the inner calls (including other
hookwrapper calls).

If "specname" is provided, it will be used instead of the function name when
matching this hook implementation to a hook specification during registration.

"""

def setattr_hookimpl_opts(func):
    setattr(
        func,
        self.project_name + "_impl",
        dict(
            hookwrapper = hookwrapper,
            optionalhook = optionalhook,
            tryfirst = tryfirst,
            trylast = trylast,
            specname = specname,
        ),
    )
    return func

if function is None:
    return setattr_hookimpl_opts
else:
    return setattr_hookimpl_opts(function)
```

如下所示,在应用实例中的两行代码实际就说明 hookspec 这个对象有一个属性 project_name,而此属性的值就是 myproject,而 hookimpl 对象也有一个属性 project_name,而此属性的值也是 myproject。

```
hookspec = pluggy.HookspecMarker("myproject")
hookimpl = pluggy.HookimplMarker("myproject")
```

这里使用 hookspec 作为装饰器作用在 myhook 方法上,其实就是调用 HookspecMarker 中的 __call__ 方法,而此方法中的 function 参数就是 myhook 方法。其他几个参数默认为 None,从 HookspecMarker 中的 __call__ 方法中可以看出,当 function 存在值时,实际是为 function 设置了一个属性,也就是为 myhook 方法设置了一个 self.project_name + "_spec",即 myproject_spec 属性的值就是代码中的这个 dict 字典。

```python
class MySpec:
    @hookspec
    def myhook(self, arg1, arg2):
        pass
```

同理，在插件定义实现接口的类中，也是同样的，我们先调用 HookimplMarker 中的 __call__ 方法，在这里此方法中 function 参数的值同样为 myhook，当 function 不为空值的时候，就是给 myhook 设置了一个 self.project_name ＋ "_impl"也就是 myhook_impl 的属性，这里的属性值同样是 dict 字典数据。

```python
class Plugin_1:
    @hookimpl
    def myhook(self, arg1, arg2):
        print("in Plugin_1.myhook()")
        return arg1 + arg2
```

至此，HookspecMarker 类和 HookimplMarker 类的源码就解析完了，这里需要注意，Python 语言中 __call__ 魔法函数是用来做装饰器的，了解这一点，那么这两个类的代码定义就很容易理解了。

15.4 如何将 PluginManager 类实例化

在解析 PlugginManager 类之前，首先再来看下一下 Pluggy 的应用实例代码。通过前面的源码解析得知，此时已经存在两个实例 hookspec 和 hookimpl，此外，MySpec 类中的 myhook 方法新增了一个 myproject_spce 的属性，属性值是一个字典，字典包含三个 key，分表是 firstresult、historic、warn_on_impl，而 Plugin_1、Plugin_2、Plugin_3 类中的 myhook 方法也同样都增加了一个 myproject_impl 属性，属性值也是一个字典，字典包含 hookwrapper、optionalhook、tryfirst、trylast、specname 五个 key。其中 Plugin_2 中 myhook myproject_impl 属性中的 hookwrapper 的值为 True，这是在类中使用装饰器时通过装饰器传入参数的方式设置的。

```python
import pluggy

hookspec = pluggy.HookspecMarker("myproject")
hookimpl = pluggy.HookimplMarker("myproject")

class MySpec:
    @hookspec
    def myhook(self, arg1, arg2):
        pass
```

```python
class Plugin_1:
    @hookimpl
    def myhook(self, arg1, arg2):
        print("in Plugin_1.myhook()")
        return arg1 + arg2

class Plugin_2:
    @hookimpl(hookwrapper = True)
    def myhook(self, arg1, arg2):
        print("in Plugin_2.myhook() before yield...")
        output = yield
        result = output.get_result()
        print("in Plugin_2.myhook() after yield...")
        print(result)

class Plugin_3:
    @hookimpl
    def myhook(self, arg1, arg2):
        print("inside Plugin_3.myhook()")
        return arg1 - arg2 + 10

pm = pluggy.PluginManager("myproject")
pm.add_hookspecs(MySpec)
pm.register(Plugin_1())
pm.register(Plugin_2())
pm.register(Plugin_3())
results = pm.hook.myhook(arg1 = 1, arg2 = 2)
print("after all run ...")
print(results)
```

下面就开始对 PluginManager 类进行实例化，实例化的对象为 pm，因为 PluginManager 类中的代码比较多，而实例化时只会调用类中的 __init__ 方法。我们先仅列出了 PluginManager 类中 __init__ 方法的源码。这里对 project_name 属性赋值，如赋值为 myproject，然后初始化几个变量，从 _name2plugin 名称看可以推断出插件的名字和插件对象的映射关系，推断出 _plugin2hookcallers 大概是插件和 hookcaller 的对应关系。目前推断不出来 _plugin_distinfo 具体功能，只能猜测是插件的某方面的信息，这个列表可用于后续存储插件。hook 则是 _HookRelay 类的一个实例，_inner_hookexec 则是一个函数，即是 _multicall 函数的别名，通过此文件的 import 部分，我们可以找到此函数在 __callers.py 文件中的定义。通过名称 trace 可以知道这是帮助打印调用追踪栈的功能，从 pluggy 模块的主要功能来看，我们可以暂时不管这个辅助功能。因此，此时要关注的主要就是 hook 和 _inner_hookexec 这两个属性，其他属性基本都是初始化操作,待后续使用时再具体分析。

```
class PluginManager:
    """Core :py:class:'.PluginManager' class which manages registration
    of plugin objects and 1:N hook calling.

    You can register new hooks by calling :py:meth:'add_hookspecs(module_or_class)
    <.PluginManager.add_hookspecs>'.
    You can register plugin objects (which contain hooks) by calling
    :py:meth:'register(plugin) <.PluginManager.register>'. The :py:class:
'.PluginManager'
    is initialized with a prefix that is searched for in the names of the dict
    of registered plugin objects.

    For debugging purposes you can call :py:meth:'.PluginManager.enable_tracing'
    which will subsequently send debug information to the trace helper.
    """

    def __init__(self, project_name):
        self.project_name = project_name
        self._name2plugin = {}
        self._plugin2hookcallers = {}
        self._plugin_distinfo = []
        self.trace = _tracing.TagTracer().get("pluginmanage")
        self.hook = _HookRelay()
        self._inner_hookexec = _multicall
```

通过上述 import 部分内容可以查看到_HookRelay 类是在_hook.py 中定义的,而此类的定义是一个空类,因此,可以知道这个类应只是为了组织存储数据的,需要后续动态增加设置属性。

```
class _HookRelay:
    """hook holder object for performing 1:N hook calls where N is the number
    of registered plugins.

    """
```

这里的_inner_hookexec 指定的是_callers.py 文件中的_multicall 函数,我们暂时先不详细解释此函数的功能。只告诉大家它是 Pluggy 模块中调用函数执行核心中的核心。

```
def _multicall(hook_name, hook_impls, caller_kwargs, firstresult):
    """Execute a call into multiple python functions/methods and return the
    result(s).
```

```python
    "caller_kwargs" comes from _HookCaller.__call__().
    """
    __tracebackhide__ = True
    results = []
    excinfo = None
    try:  # run impl and wrapper setup functions in a loop
        teardowns = []
        try:
            for hook_impl in reversed(hook_impls):
                try:
                    args = [caller_kwargs[argname] for argname in hook_impl.argnames]
                except KeyError:
                    for argname in hook_impl.argnames:
                        if argname not in caller_kwargs:
                            raise HookCallError(
                                f"hook call must provide argument {argname!r}"
                            )

                if hook_impl.hookwrapper:
                    try:
                        gen = hook_impl.function(*args)
                        next(gen)  # first yield
                        teardowns.append(gen)
                    except StopIteration:
                        _raise_wrapfail(gen, "did not yield")
                else:
                    res = hook_impl.function(*args)
                    if res is not None:
                        results.append(res)
                        if firstresult:  # halt further impl calls
                            break
        except BaseException:
            excinfo = sys.exc_info()
    finally:
        if firstresult:  # first result hooks return a single value
            outcome = _Result(results[0] if results else None, excinfo)
        else:
            outcome = _Result(results, excinfo)

        # run all wrapper post-yield blocks
        for gen in reversed(teardowns):
            try:
                gen.send(outcome)
```

```
            _raise_wrapfail(gen, "has second yield")
        except StopIteration:
            pass

    return outcome.get_result()
```

15.5 为 add_hookspecs 增加自定义的接口类

在实践中，add_hookspecs 的应用代码将 MySpec 类会作为参数传递给 add_hookspecs 方法。

pm.add_hookspecs(MySpec)

我们看 add_hookspecs 方法的定义，这里形参 module_or_class 即是 MySpec 类，因此这里的 for 循环相当于是遍历 MySpec 类的方法和属性。从最外层的 for 循环和与 for 循环同层级的 if 语句可以看出，for 循环就是通过对传进来的 MySpec 类的方法和属性进行遍历。符合要求的就放入 names 列表，如果判断 names 为空，即没有找到符合要求的方法或属性，则抛出一个异常。那到底什么样的才是符合要求的，我们继续分析。

```
def add_hookspecs(self, module_or_class):
    """add new hook specifications defined in the given ``module_or_class``.
    Functions are recognized if they have been decorated accordingly."""
    names = []
    for name in dir(module_or_class):
        spec_opts = self.parse_hookspec_opts(module_or_class, name)
        if spec_opts is not None:
            hc = getattr(self.hook, name, None)
            if hc is None:
                hc = _HookCaller(name, self._hookexec, module_or_class, spec_opts)
                setattr(self.hook, name, hc)
            else:
                # plugins registered this hook without knowing the spec
                hc.set_specification(module_or_class, spec_opts)
                for hookfunction in hc.get_hookimpls():
                    self._verify_hook(hc, hookfunction)
            names.append(name)

    if not names:
        raise ValueError(
            f"did not find any {self.project_name!r} hooks in {module_or_class!
```

r}"
)

我们知道 spec_opts 变量是通过 parse_hookspec_opts 方法来获取值的,而从此方法的实现看出,其先根据方法或属性名获取方法或属性对象,然后再将该对象 myproject_spec 属性的值返回。通过前面解析 HookspecMarker 类我们知道,定义 MySpec 类中的 myhook 方法时使用了一个装饰器,而此装饰器就是增加 myproject_spec 属性的。至此,就得出 parse_hookspec_opts 方法作用就是获取对象的 myproject_spec 属性的值。如果没有此属性时,则返回 None。

```
def parse_hookspec_opts(self, module_or_class, name):
    method = getattr(module_or_class, name)
    return getattr(method, self.project_name + "_spec", None)
```

再返回到 add_hookspecs 方法,这里可以看到,如果 spec_opts 是 None 就直接跳过了,只有 spec_opts 不是 None 时才会做进一步的处理。换句话说,MySpec 的所有方法和属性中,只有被 hookspec 装饰器装饰了的函数,才进行进一步处理,而在前面也已经分析过,MySpec 类中 myhook 方法使用了装饰器后,就拥有了 myproject_spec 属性,且属性值是一个字典。其字典中有 firstresult、historic、warn_on_impl 三个 key。下面的 hc 变量会从 self.hook 对象中取 myhook 属性。在 PluginManager 类初始化的时候,我们知道,self.hook 被初始化为_HookRelay 类的一个实例,但此类没有定义任何属性和方法,因此这里的 hc 显然为 None。接下来当我们看到 hc 为 None 时,hc 给了初始化为_HookCaller 类一个实例,从_HookCaller 类的初始化实现代码中可以看到,这里的 name 就是注册的接口名 myhook,这里的_hookexec 就是 PluginManager 类中的_inner_hookexec,也就是_multicall 函数,而 specmodule_or_class 则是传递进来的 MySpec 类,显然不是 None,因此又继续调用了 specification 方法。

```
class _HookCaller:
    def __init__(self, name, hook_execute, specmodule_or_class = None, spec_opts = None):
        self.name = name
        self._wrappers = []
        self._nonwrappers = []
        self._hookexec = hook_execute
        self._call_history = None
        self.spec = None
        if specmodule_or_class is not None:
            assert spec_opts is not None
            self.set_specification(specmodule_or_class, spec_opts)
```

在 set_specification 方法定义的代码中,给_HookCaller 类的 spec 属性赋值了一个 HookSpec 类的实例化对象,会并判断 spec_opts 中 historic 是否为 True。如果为 True,则将_HookCaller 类的_call_history 属性初始化为空列表。

```python
def set_specification(self, specmodule_or_class, spec_opts):
    assert not self.has_spec()
    self.spec = HookSpec(specmodule_or_class, self.name, spec_opts)
    if spec_opts.get("historic"):
        self._call_history = []
```

从 HookSpec 类的定义中看出，这里有一个原子类型的类，且此类中的 namespace 是 MySpec 类，name 是定义的接口名 myhook，function 是 myhook 方法对象，opts 是包含 firstresult，historic，warn_on_impl 的三个 key 字典，argnames 和 kwargnames 是 myhook 函数的参数，warn_on_impl 则是 opts 中的 warn_on_impl 属性值。

```python
class HookSpec:
    def __init__(self, namespace, name, opts):
        self.namespace = namespace
        self.function = function = getattr(namespace, name)
        self.name = name
        self.argnames, self.kwargnames = varnames(function)
        self.opts = opts
        self.warn_on_impl = opts.get("warn_on_impl")
```

我们再回到 add_hookspecs 方法中，看到此时 hc 已经赋值了，hc 是 _HookCaller 类的一个实例对象，然后，其下面一行代码是 self.hook 设置的一个属性和值，是给 _HookRelay 类的实例对象 self.hook 设置的属性 myhook，值为 hc 对象。

```python
setattr(self.hook, name, hc)
```

当然，如果 hc 不为空，则会直接更新属性。

至此，add_hookspecs 方法就分析完成了。

15.6　register 注册插件源码解析

首先，我们看应用中注册的代码，这里调用了 register 函数，传入了实现接口的插件类。

```python
pm.register(Plugin_1())
```

register 方法定义如下所示。这里首先获取了插件的名称，在注册插件的时候，可以通过 name 形参指定插件的名称，如果没有指定，则通过 get_canonical_name 方法获取。

```python
def register(self, plugin, name=None):
    """Register a plugin and return its canonical name or ``None`` if the name
    is blocked from registering.  Raise a :py:class:`ValueError` if the plugin
    is already registered."""
    plugin_name = name or self.get_canonical_name(plugin)
```

```python
        if plugin_name in self._name2plugin or plugin in self._plugin2hookcallers:
            if self._name2plugin.get(plugin_name, -1) is None:
                return  # blocked plugin, return None to indicate no registration
            raise ValueError(
                "Plugin already registered: %s=%s\n%s"
                % (plugin_name, plugin, self._name2plugin)
            )

        # XXX if an error happens we should make sure no state has been
        # changed at point of return
        self._name2plugin[plugin_name] = plugin

        # register matching hook implementations of the plugin
        self._plugin2hookcallers[plugin] = hookcallers = []
        for name in dir(plugin):
            hookimpl_opts = self.parse_hookimpl_opts(plugin, name)
            if hookimpl_opts is not None:
                normalize_hookimpl_opts(hookimpl_opts)
                method = getattr(plugin, name)
                hookimpl = HookImpl(plugin, plugin_name, method, hookimpl_opts)
                name = hookimpl_opts.get("specname") or name
                hook = getattr(self.hook, name, None)
                if hook is None:
                    hook = _HookCaller(name, self._hookexec)
                    setattr(self.hook, name, hook)
                elif hook.has_spec():
                    self._verify_hook(hook, hookimpl)
                    hook._maybe_apply_history(hookimpl)
                hook._add_hookimpl(hookimpl)
                hookcallers.append(hook)
        return plugin_name
```

get_canonical_name 方法的定义如下所示，可以看出，这里首先获取了插件类的__name__属性值。如果没有此属性，则会返回插件类的 id 作为插件的名称。

```python
    def get_canonical_name(self, plugin):
        """Return canonical name for a plugin object. Note that a plugin
        may be registered under a different name which was specified
        by the caller of :py:meth:`register(plugin, name) <.PluginManager.register>`.
        To obtain the name of an registered plugin use :py:meth:`get_name(plugin)
        <.PluginManager.get_name>` instead."""
        return getattr(plugin, "__name__", None) or str(id(plugin))
```

我们这回看 register 方法，如果插件名称已经存在于_name2plugin 属性中或者插件类对象已经存在于_plugin2hookcallers 属性中了，说明插件已经注册过了，则报错提醒。如果没有注册,那么可以看到,_name2plugin 属性果然是存放了插件名称和插件类对象的一个属性。

```
self._name2plugin[plugin_name] = plugin
```

接下来一行代码则说明 plugin2hookcallers 是存放插件和 hookcallers 映射关系的属性,因为一个插件完全可以存在多个 hookcallers,因此,这里初始化为了一个空的列表。

```
self._plugin2hookcallers[plugin] = hookcallers = []
```

我们看到,for 循环部分和 add_hookspecs 中的解析接口类属性的代码是类似的,即对 HookimplMarker 类的所有方法和属性进行循环遍历,针对每一个方法或属性,首先解析 hookimpl_opts 的值。从 parse_hookimpl_opts 的定义代码中看出,如果没有被修饰说明不是要找的方法,直接返回进入下一个循环,如果是,则再获取 myproject_impl 属性。同样,通过对前面对 HookimplMarker 类的初始化代码分析得知,myproject_impl 属性的值也是一个字典,而且此字典包含 hookwrapper、optionalhook、tryfirst、trylast、specname 五个 key,我们这里 res 里存的就是由这五个 key 组成的字典。

```
def parse_hookimpl_opts(self, plugin, name):
    method = getattr(plugin, name)
    if not inspect.isroutine(method):
        return
    try:
        res = getattr(method, self.project_name + "_impl", None)
    except Exception:
        res = {}
    if res is not None and not isinstance(res, dict):
        # false positive
        res = None
    return res
```

我们回到 register 方法中,当 hookimpl_opts 不为空时,此时的方法是被 hookimpl 装饰的,也就是我们所要寻找的 myhook 实现方法。我们再对 hookimpl_opts 进行一次规范化处理,实现代码如下所示。

```
def normalize_hookimpl_opts(opts):
    opts.setdefault("tryfirst", False)
    opts.setdefault("trylast", False)
    opts.setdefault("hookwrapper", False)
    opts.setdefault("optionalhook", False)
    opts.setdefault("specname", None)
```

接下来中,我们看 hookimpl 为 HookImpl 类的一个实例对象,从 HookImpl 类的代码中,可以看出,这里 HookImpl 类和 HookSpec 类是相似的,主要用于存放插件类属性,比如这里就存放了插件类中的方法、方法参数、插件、插件名称以及修饰的参数等。出现的__repr__方法是 Python 语言中的常用的一个魔法函数,用于打印对象时在控制台显示的内容。

```
class HookImpl:
    def __init__(self, plugin, plugin_name, function, hook_impl_opts):
        self.function = function
        self.argnames, self.kwargnames = varnames(self.function)
        self.plugin = plugin
        self.opts = hook_impl_opts
        self.plugin_name = plugin_name
        self.__dict__.update(hook_impl_opts)

    def __repr__(self):
        return f"<HookImpl plugin_name={self.plugin_name!r}, plugin={self.plugin!r}>"
```

再次回到 register 方法。对于 name 的取值情况,我们首先看 hookimpl_opts,如果 hookimpl_opts 字典中有 specname,则取其值,否则则直接取方法名,如从当前类的 self.hook 对象中获取 myhook 的值,由于在分析 add_hookspecs 时,曾经分析过,self.hook 中增加了名为 myhook 的属性,myhook 属性的值为 add_hookspecs 方法中的 hc_HookCaller 类的对象,因此,这里 hook 可以获取 add_hookspecs 中的 hc。之后,我们通过_add_hookimpl 方法继续对 hook 增加属性配置。

```
hook._add_hookimpl(hookimpl)
```

如下所示,从_add_hookimpl 方法中可以看到,如果 hookimpl 的 trylast 为 True,则会在方法列表中将当前的 hookimpl 插入到第一个位置,这就是为什么在插件定义中如果使用了 trylast=True 时,插件会最后一个执行,因为这些方法是先进后出的队列。同样如果设置了 tryfirst=True,则会将当前 hookimpl 使用的 append 插入到最后,这样最后一个就要先执行了。此外 hookimpl 的 hookwrapper 若为 True,则表示此方法中有 yield 关键字,则此时要将 hookimpl 放入_wrappers 属性中,而若为 False,则直接将 hookimpl 放入_nonwrappers 属性中。

```
def _add_hookimpl(self, hookimpl):
    """Add an implementation to the callback chain."""
    if hookimpl.hookwrapper:
        methods = self._wrappers
    else:
        methods = self._nonwrappers

    if hookimpl.trylast:
        methods.insert(0, hookimpl)
    elif hookimpl.tryfirst:
```

```
            methods.append(hookimpl)
     else:
        # find last non-tryfirst method
        i = len(methods) - 1
        while i >= 0 and methods[i].tryfirst:
            i -= 1
        methods.insert(i + 1, hookimpl)
```

最后,将 hook 放入_plugin2hookcallers 属性的 plugin 类对象作为 key 的值中。至此,register 方法就解析完成了。

15.7 hook 函数调用执行过程分析

首先,我们看一下 Pluggy 应用代码中执行 hook 函数调用的代码。在分析 add_hookspecs 方法的时候,我们曾经分析过,pm 有一个属性是 hook,而 hook 实质上是_HookRelay 类的实例对象,但是_HookRelay 类是一个空类,因为在 add_hookspecs 中对 hook 这个对象设置了一个属性,属性名是 myhook,属性的值是 hc,而 hc 实质上是_HookCaller 类的一个实例,即 pm.hook.myhook 就是_HookCaller 类的一个实例,那么这里是将实例当作函数调用,很显然,在 Python 中,这种用法实质上是在调用_HookCaller 类中的__call__魔法函数。

```
results = pm.hook.myhook(arg1 = 1, arg2 = 2)
```

接下来进入_HookCaller 类中,确实可以找到__call__魔法函数,这里可以看到,首先获取 firstresult 是否设置为 True,如果没有设置,则直接将 firstresult 设置为 False,然后调用_hookexec 方法,而在_HookCaller 类的初始化函数中,可以看出_hookexec 方法就是传递进来的 PluginManager 类的_inner_hookexec 属性,即_multicall 函数。

```
def __call__(self, *args, **kwargs):
    if args:
        raise TypeError("hook calling supports only keyword arguments")
    assert not self.is_historic()

    # This is written to avoid expensive operations when not needed.
    if self.spec:
        for argname in self.spec.argnames:
            if argname not in kwargs:
                notincall = tuple(set(self.spec.argnames) - kwargs.keys())
                warnings.warn(
                    "Argument(s) {} which are declared in the hookspec "
                    "can not be found in this hook call".format(notincall),
```

```
                    stacklevel = 2,
                )
                break

        firstresult = self.spec.opts.get("firstresult")
    else:
        firstresult = False

    return self._hookexec(self.name, self.get_hookimpls(), kwargs, firstresult)
```

因此，pm.hook.myhook 实质上就是调用了 _multicall 函数，在前面也曾经说过，_multicall 函数是整个 Pluggy 模块调用执行的核心。下面详细介绍此函数的实现，我们看 _multicall 的函数代码，在 for 循环时，将 hook_impls 使用 reverse 进行了反转，前面讲述，在添加执行函数时好像使用了先进后出队列，显然这里并没有使用队列的数据结构，而是使用了列表，只对列表进行了反转，在每个循环中判断 hookwrapper 的值是否为 True。如果 hookwrapper 为 True，则表示方法中有 yield，此时就需要将 yield 之后的调用提前存入 teardowns 列表，同时执行所有的函数。此外，在判断 firstresult 时，如果 firstresult 为 True，那么当有一个结果时就会停止执行。在 finally 部分我们看到，这里将 teardowns 进行反转后再依次执行调用，这就做到了在有多个插件类的方法中使用 yield 时，先注册插件类中 yield 之后的代码是后执行的，而这个功能对于 Pytest 中的 teardown 很有用处。

```
def _multicall(hook_name, hook_impls, caller_kwargs, firstresult):
    """Execute a call into multiple python functions/methods and return the
    result(s).

    "caller_kwargs" comes from _HookCaller.__call__().
    """
    __tracebackhide__ = True
    results = []
    excinfo = None
    try:   # run impl and wrapper setup functions in a loop
        teardowns = []
        try:
            for hook_impl in reversed(hook_impls):
                try:
                    args = [caller_kwargs[argname] for argname in hook_impl.argnames]
                except KeyError:
                    for argname in hook_impl.argnames:
                        if argname not in caller_kwargs:
                            raise HookCallError(
                                f"hook call must provide argument {argname!r}"
                            )
```

```python
            if hook_impl.hookwrapper:
                try:
                    gen = hook_impl.function(*args)
                    next(gen)    # first yield
                    teardowns.append(gen)
                except StopIteration:
                    _raise_wrapfail(gen, "did not yield")
            else:
                res = hook_impl.function(*args)
                if res is not None:
                    results.append(res)
                    if firstresult:    # halt further impl calls
                        break
    except BaseException:
        excinfo = sys.exc_info()
    finally:
        if firstresult:    # first result hooks return a single value
            outcome = _Result(results[0] if results else None, excinfo)
        else:
            outcome = _Result(results, excinfo)

        # run all wrapper post-yield blocks
        for gen in reversed(teardowns):
            try:
                gen.send(outcome)
                _raise_wrapfail(gen, "has second yield")
            except StopIteration:
                pass

        return outcome.get_result()
```

至此，hook 函数调用执行的源码就解析完成了。

15.8　PluginManager 类的其他功能

前面 Pluggy 的主要功能的源码我们在前面章节中基本解读完了，这里将 PluginManager 类中其他功能的代码简要解读一下。PluginManager 类不但提供了注册插件的功能，还提供了取消注册插件的功能，注册的代码解读完成后再来看取消注册的代码就简单多了。简单点说，取消注册就是 _name2plugin 在属性中删除对应的插件，然后将 _plugin2hookcallers 属性中对应插件的所有注册方法去除即可，当然，在这个过程中，还会涉及判断是否存在此插件等问题。

```python
    def unregister(self, plugin=None, name=None):
```

```python
"""unregister a plugin object and all its contained hook implementations
from internal data structures."""
if name is None:
    assert plugin is not None, "one of name or plugin needs to be specified"
    name = self.get_name(plugin)

if plugin is None:
    plugin = self.get_plugin(name)

# if self._name2plugin[name] == None registration was blocked: ignore
if self._name2plugin.get(name):
    del self._name2plugin[name]

for hookcaller in self._plugin2hookcallers.pop(plugin, []):
    hookcaller._remove_plugin(plugin)
```

PluginManager 提供了设置阻塞的方法，设置阻塞的代码很简单，直接调用取消注册的方法实现，然后，将此插件名对应的值设置为 None 即可。

```python
def set_blocked(self, name):
    """block registrations of the given name, unregister if already registered."""
    self.unregister(name=name)
    self._name2plugin[name] = None
```

下面是判断是否阻塞的代码实现情况，实质上就是判断插件名是否在_name2plugin 属性中，以及当_name2plugin 属性中存在插件名时需要进一步判断此插件名对应的插件对象是否为 None。

```python
def is_blocked(self, name):
    """return ''True'' if the given plugin name is blocked."""
    return name in self._name2plugin and self._name2plugin[name] is None
```

获取插件列表的代码时，直接将_plugin2hookcallers 属性转换为集合，然后返回，如下所示。

```python
def get_plugins(self):
    """return the set of registered plugins."""
    return set(self._plugin2hookcallers)
```

如下代码可判断插件是否注册，即判断插件类对象是否在_plugin2hookcallers 属性中。

```python
def is_registered(self, plugin):
    """Return ''True'' if the plugin is already registered."""
    return plugin in self._plugin2hookcallers
```

如下所示内容为根据插件名获取插件对象。这里直接从_name2plugin属性中根据名称获取即可,因为_name2plugin是一个字典格式数据。

```python
def get_plugin(self, name):
    """Return a plugin or "None" for the given name."""
    return self._name2plugin.get(name)
```

想判断是否有插件,直接调用根据名称获取插件对象,只要返回的插件对象不为None即可。

```python
def has_plugin(self, name):
    """Return "True" if a plugin with the given name is registered."""
    return self.get_plugin(name) is not None
```

我们可以通过指定group名称的方式自动注册插件,只要当前的环境中其他包中有通过setuptools的方式定义了此group名称的模块,就可以自动进行注册,在著名的自动化测试框架Pytest中,就使用了这个函数。对group为"Pytest11"的插件进行自动注册,使用起来非常方便,用户可以通过自定义Pytest11模块来对Pytest做功能增强,而不需要修改Pytest的源码。

```python
def load_setuptools_entrypoints(self, group, name=None):
    """Load modules from querying the specified setuptools "group".

    :param str group: entry point group to load plugins
    :param str name: if given, loads only plugins with the given "name".
    :rtype: int
    :return: return the number of loaded plugins by this call.
    """
    count = 0
    for dist in list(importlib_metadata.distributions()):
        for ep in dist.entry_points:
            if (
                ep.group != group
                or (name is not None and ep.name != name)
                # already registered
                or self.get_plugin(ep.name)
                or self.is_blocked(ep.name)
            ):
                continue
            plugin = ep.load()
            self.register(plugin, name=ep.name)
            self._plugin_distinfo.append((plugin, DistFacade(dist)))
            count += 1
    return count
```

PluginManager类还有几个简单的方法,这里就不再一一分析了,至此对Pluggy的源码分析就全部完成了。